Classic Chevies
of the
'50s
A Maintenance &Repair Manual

Orest Lazarowich

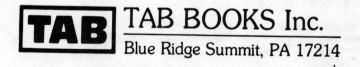
TAB TAB BOOKS Inc.
Blue Ridge Summit, PA 17214

629.28722

FIRST EDITION

FIRST PRINTING

10-87 BT 2200

Copyright © 1986 by TAB BOOKS Inc.

Printed in the United States of America

Library of Congress Cataloging in Publication Data

Lazarowich, Orest.
    Classic Chevies of the '50s.

    Includes index.
       1. Automobiles—Conservation and restoration.
2. Chevrolet automobile. I. Title. II. Title:
Classic Chevies of the fifties.
TL152.2.L38   1985   629.28'722   85-22189
ISBN 0-8306-2115-6 (pbk.)

Front cover photograph courtesy of John Radesh.
Illustrations and appendices courtesy of Chevrolet Motor Division, General Motors Corporation.

Contents

tools a mechanic does but at least consider a similar line. A little less finish and a chrome-vanadium alloy steel with a forged design are the best bet. It takes a number of years and dollars to build up a suitable tool selection, but with the cost spread over a number of years, it becomes easier to handle. We all want the right tool at the right time for the right job. Sometimes this is not possible at the start. Don't buy specialized tools at the beginning. Start with a good, basic repair kit.

A set of combination wrenches that have a box end and an open end are a good place to start. Two adjustable wrenches (a 6 or 8 inch and a 10 or 12 inch) will make good companions for the combination wrenches so that you can remove nuts from bolts.

Screwdrivers are next. Look around at what you already have. A good assortment is possible with the replaceable bit set that is stored in the handle. The most common type ends are the flat and the Phillips. Buy one heavy-duty screwdriver so that you can use it as a pry bar.

For pliers, the combination, adjustable jaw and vice-grips (which you probably already have) will do.

A ball-peen hammer, a 16-ounce hammer, a 1-inch cold chisel, and a tapered punch will often be of help. A hacksaw is a valuable cutting tool. Buy one with an adjustable handle so different blade lengths can be used. Use a flat file to remove burrs and flashings.

Buy one spark-plug socket of the size to fit and a style that a box end wrench can be used to turn it.

For measuring tools, use a steel rule or push pull tape. A flat blade feeler gauge and a spark plug wire type of gauge will do. A good-quality tire pressure gauge will help.

You will be able to save money by doing your own oil changes and grease jobs, but you need some special equipment. The most important item is the oil filter wrench. A grease gun is next; buy the one that takes the grease cartridge. You will also need a set of car stands of suitable strength to support the car, a drain pan for the used oil and a combination funnel, and a can opener. Check the local regulations for disposing of used oil. As you become more proficient in tool use, you might want to add a ratchet wrench set with suitable drivers and sockets. These are bought with a box so it help to keep the pieces together. Expand the punch line to include a center punch, drift punch, and a set of pin punches. Study the tool catalogs and let your family members know that you appreciate tools. You know what you need so tell them.

For portable equipment a 3/8 variable speed drill is invaluable,

Introduction

Those fabulous mid-'50s.

The 1955-1956-1957 Chevrolet product line is proving to be the best all-around bargain in the postwar field. You can still buy one at a fair price and keep it running with very little effort on your part. Parts are available either through NOS or Jobber, and the supply has not been exhausted. New sheet metal can be formed and, with a few transplants, you can put a car together that is as sharp as it was the day it rolled off the assembly line.

The 1955 model was chosen as the pace car for the Indy 500. Styling and engineering left 1954 so far behind it was the dawning of a new era. Pick a hardtop with two-tone paint and the extra chrome and you have the start of a great style. The optional 265 V-8 develops 162 horsepower, and the six cylinder has become history. Other engineering changes included ball joint front suspension, open driveshaft and Hotchkiss rear end, and a 12-volt electrical system.

For 1956, the styling changes a bit but it remains a great-looking car. Directional turn signals are standard equipment, and seat belts with shoulder harness are offered as options. In the engineering department, the lowly six cylinder is down to one model while the V-8 goes on to become the high point in Chevy go-power. There is 162 horsepower for standard shift cars, 170 horsepower for the automatic, and a special 205 horsepower engine.

And now the big one: 1957. Styling changes in the front end

and a very slight fin at the rear make this the most popular Chevrolet of the '50s. The wheel size is reduced to 14 inches and a low pressure tire is used to bring the overall height down. The 1957 Bel Air Sport Coupe is one fine-looking car. The V-8 engine options will give you just about anything you want in high-performance power. Plenty of room inside, good trunk space, and an economical engine make these cars good, daily transportation.

This book presents a preventive maintenance approach so that you, as a driveway mechanic, can keep your car running safely and economically and derive satisfaction from one of Chevrolet's best. The times are certainly showing higher car prices and smaller car values, so let's roll those nifty mid-'50s into the future.

Chapt

The Driveway Mecha

You can do mechanical work on your driveway and be succe about it. However, if you are going to do service that requires than a couple of hours of work, it's better to have a roof overh You will need storage for parts, supplies, tools, and equipment. driveway is not a good place to advertise your automotive sk unless you are having a garage sale. If you can work in a carp organize the wall area for maximum storage. You might consi renting space at a local garage to complete a service proced A single-car home garage is pretty cramped, but at least you have protection from the elements.

If the winter season tends to be one of cold and snow, insu for comfort. That garage door sure loses a lot of heat. Lightin best with one or two fluorescent fixtures at the ceiling and a d ped one over the work area. See that there are enough wall pl so that extension cords are not a hazard. Paint the walls and ing a light color to obtain best reflection. See that the floor is cl and paint it a suitable color. Make some provision for drainag ice and snow are a winter problem. The larger garage you h the more room for storage but the greater the chance of sto garbage. I know; the minute you throw something away is w you need it.

HAND TOOL SELECTION

Buy good-quality tools. You probably don't need the same ra

for drilling, buffing, grinding, and polishing. Buy high-speed drill bits of the size you need and not the drill size assortment.

CARE OF TOOLS AND EQUIPMENT

You probably practice good housekeeping at work so continue it into your personal work place. Store your tools so they can be easily located. A pegboard is a good way to group them. A tool box is great if you have the money. Construct a workbench and, if space is at a premium, make the bench portable or make it so that you can fold it against the wall.

A vise is a terrific holding tool so mount it solidly even on the wall. Storage space for oils, filters and parts can be located on the walls nearer the ceiling. This is your place so make it pleasant and comfortable to work in.

Wipe dirty tools clean before you put them away. A slightly oily cloth will do the job and help prevent the tools from rusting. Use tools that are in good condition. Striking tools such as punches and chisels can mushroom out at the ends after a period of use. Dress these ends on a grinder or a file to eliminate the possibility of tool chips breaking and flying.

Check your hammer handle to see that it is tight in the head. See that the striking surface is smooth and strike the work with the full, flat face. Glancing blows can chip the hammer face.

Wrenches must be the exact size for the bolt to prevent slippage. Six-point wrenches might help on rounded-off bolts. Pull on a wrench; don't push. If you must push, use the base of your palm to prevent "barking" your knuckles. When using an adjustable wrench, snug it up tight and apply the pulling force to the stationary jaw side of the wrench. Which wrench to use is something best learned by experience, but you will not find it difficult. Tool catalogs often have good information regarding tool use.

Screwdrivers are basically intended for tightening or loosening screws and if you do use them as pry bars you will break the tips and bend the shanks. To service a flat tip, square the tip and grind the sides almost parallel. Keep the tip cool when grinding or else the heat will draw the temper from the tip and it will become soft. The tip should fit the slot snugly and not be any wider than the screw. If the sides taper toward the end, the screwdriver will raise out of the slot when the twisting force is applied. A Phillips tip can be serviced by maintaining the same tip angle but the tip must fit properly in the screw slots.

The file and the hacksaw are probably the most abused of the cutting tools. Never use a file without a handle. You might drive the tang of the file into your hand. Files are available in single or double cut, and in a variety of shapes to suit the job being done. Clamp the work in the vise with the top of the work level with your elbow. If the part has a finished surface, use a few layers of masking tape to prevent marking. When taking a heavy cut or rough filing, use a stroke of about 50 degrees and use enough pressure to keep the file cutting. For finish filing, use a fine-cut file and light pressure. Release the pressure on the back stroke and raise the file to prevent dulling the cutting edge of the teeth.

Draw filing is a method used to produce a smooth finish on the work. A single cut file is drawn crosswise on the work, with pressure applied on both forward and backward strokes. Keep the file clean to prevent small particles from scratching the work. A narrow block of wood pushed across the file, parallel to the teeth, will clean them. Rubbing chalk into the file teeth might prevent scratching. Keep the file clean of cuttings. Store files in a rack to protect their cutting edges.

When using a hacksaw, place the blades in so that the teeth point away from the handle. This means that the cutting action is on the forward stroke. Use sufficient pressure so that the teeth will cut, and use the full length of the blade at about 40 strokes per minute. A blade with 18 teeth per inch is good for all-purpose work. Two or more teeth should be in contact with the work at all times.

Pliers are not to be used for tightening or loosening nuts or bolts. They also are not to be used for or on brass fittings. Don't use them on hardened surfaces because it dulls the teeth and the jaws lose their grip. Use pliers only when no other tool will do the job.

In addition to oils and greases, you will need some other supplies: wiping rags or paper towels to clean up spills; a metal, air-tight container for disposing of oily rags; air, oil, and gas filters; electrical and masking tape; cleaning solvent; a spray bottle you can fill with rust penetrating oil; assorted nuts, bolts, washers (both flat and lock), and cotter pins.

There are lots of new gasket sealers on the market. Check on chemical welding supplies such as epoxy. A soldering gun and solder for making up good electrical joints and a bench grinder with a buffing wheel for cleaning parts are real timesavers.

If you are going to buy test equipment, make sure it will work on your car. You should consider a torque wrench if you want to get into the real take-apart-and-reassemble area of service work. Basically, you are only limited by what you want to do and the tools you have for the job. Get in to this slowly and enjoy the jobs you can do. If you remove the faulty part and replace it with a rebuilt unit that you have bought as an exchange, that is fine. You've got the car running, and that's what it's all about.

SAFETY

You must practice personal and property safety in your work area. The service and maintenance of the automobile presents conditions that can be a safety hazard to life and limb. An accident can turn a pleasant experience into tragedy. The possibility of a fire is one of the greatest dangers. A shop without a working foam or carbon-dioxide extinguisher is not a safe place. Make an extinguisher your first purchase, and know how to use it.

Store gasoline only in approved containers. You already have a gas tank in the car. How much more do you need? Label any other containers that hold oil, antifreeze, solvents, etc. Do not use gasoline to wash parts or tools; use solvent. Dispose of oily rags.

Do not operate a car engine indoors. Have a flexible exhaust tube to carry the fumes outside. Carbon monoxide is a killer. This gas is odorless, colorless, and tasteless, and it is a product of combustion. Don't operate the car engine in a closed garage. In three minutes, a small garage can become filled with enough carbon monoxide to kill you.

Take care when removing the radiator cap from a hot or boiling radiator. The hot coolant can shoot up directly at you and painful burns can result. If you cannot comfortably keep your hand on the top of the radiator, leave the cap on until the radiator cools down. Be careful around the engine when the cap until the radiator fan is turning. Don't get your fingers in the way of the blades. Many times you will have to make adjustments with the engine running. Watch out for the hot exhaust system. At the same time, watch out for the fan belts because they can pull in a piece of loose clothing.

Battery acid is corrosive. Use rubber-backed carpet remnants as fender covers when servicing the battery. Neutralize skin or clothing with baking soda and water. Do not smoke around the battery because an explosion could result. Hydrogen gas is generated as a natural result of battery action, and it is possible to ignite this gas with a spark or flame. Get in the habit of disconnecting the battery ground strap when working under the hood. Don't spark a battery with a screwdriver or a pair of pliers to see if it has any charge. You might end up with no battery at all.

Don't trust any hydraulic or mechanical jack support an automobile, and never get under a car when it is only on a jack. Use car stands of the rated weight load to support the car. When two wheels are off the floor, block the other wheels to prevent the vehicle from rolling.

Have a first-aid kit handy so that you can look after burns and cuts immediately. Many parts of the engine are hot so learn to avoid personal contract with them. Keep your hands away from the fan; it cuts meat readily. Do as much work as you can on the engine while it is not running.

When grinding or buffing, use a face shield or safety glasses. They haven't yet made a glass eye that you can see through. Make sure that electrical equipment is properly grounded. Don't break the ground prong off or you will complete the circuit. Check all cords for chaffing or breaks and have them repaired. Service work can be very satisfying, but it can also become a hazard if procedures are not followed in a responsible manner. There is absolutely no place for poor workmanship. You endanger your life and the lives of others with slipshod work. If you are not willing to do the best work on your own vehicle, then leave it to the commercial shops before you become a statistic.

Be safety alert all the time and consider that accidents can happen to you. They don't just happen to someone else. With just a bit of planning and foresight, you can enjoy service work. Remember the ABCs: Always Be Careful.

Make up an emergency kit for the road. See that you have some basic tools, spare headlight and bulbs, booster cables, and fan belt. A fire extinguisher and a set of flares is a good investment.

IDENTIFYING YOUR CAR

You must present detailed information when ordering parts. It is not enough to know the year, model, and type of engine. See Figs. 1-1 through 1-9 and Table 1-1. Locate the indentification plate that lists the year, model, serial number, engine number, and paint code. Additional unit serial numbers will help you to identify component parts. You can buy parts from the automotive dealership, jobbers, or the local wrecking yard. If you cannot buy the parts to rebuild a certain component, you can buy an exchange unit instead. This is a good way to get started. As you gain more experience, you can do more of your own work.

Introduction

Those fabulous mid-'50s.

The 1955-1956-1957 Chevrolet product line is proving to be the best all-around bargain in the postwar field. You can still buy one at a fair price and keep it running with very little effort on your part. Parts are available either through NOS or Jobber, and the supply has not been exhausted. New sheet metal can be formed and, with a few transplants, you can put a car together that is as sharp as it was the day it rolled off the assembly line.

The 1955 model was chosen as the pace car for the Indy 500. Styling and engineering left 1954 so far behind it was the dawning of a new era. Pick a hardtop with two-tone paint and the extra chrome and you have the start of a great style. The optional 265 V-8 develops 162 horsepower, and the six cylinder has become history. Other engineering changes included ball joint front suspension, open driveshaft and Hotchkiss rear end, and a 12-volt electrical system.

For 1956, the styling changes a bit but it remains a great-looking car. Directional turn signals are standard equipment, and seat belts with shoulder harness are offered as options. In the engineering department, the lowly six cylinder is down to one model while the V-8 goes on to become the high point in Chevy go-power. There is 162 horsepower for standard shift cars, 170 horsepower for the automatic, and a special 205 horsepower engine.

And now the big one: 1957. Styling changes in the front end

and a very slight fin at the rear make this the most popular Chevrolet of the '50s. The wheel size is reduced to 14 inches and a low pressure tire is used to bring the overall height down. The 1957 Bel Air Sport Coupe is one fine-looking car. The V-8 engine options will give you just about anything you want in high-performance power. Plenty of room inside, good trunk space, and an economical engine make these cars good, daily transportation.

This book presents a preventive maintenance approach so that you, as a driveway mechanic, can keep your car running safely and economically and derive satisfaction from one of Chevrolet's best. The times are certainly showing higher car prices and smaller car values, so let's roll those nifty mid-'50s into the future.

Chapter 1

The Driveway Mechanic

You can do mechanical work on your driveway and be successful about it. However, if you are going to do service that requires more than a couple of hours of work, it's better to have a roof overhead. You will need storage for parts, supplies, tools, and equipment. The driveway is not a good place to advertise your automotive skills, unless you are having a garage sale. If you can work in a carport, organize the wall area for maximum storage. You might consider renting space at a local garage to complete a service procedure. A single-car home garage is pretty cramped, but at least you will have protection from the elements.

If the winter season tends to be one of cold and snow, insulate for comfort. That garage door sure loses a lot of heat. Lighting is best with one or two fluorescent fixtures at the ceiling and a dropped one over the work area. See that there are enough wall plugs so that extension cords are not a hazard. Paint the walls and ceiling a light color to obtain best reflection. See that the floor is clean and paint it a suitable color. Make some provision for drainage if ice and snow are a winter problem. The larger garage you have the more room for storage but the greater the chance of storing garbage. I know; the minute you throw something away is when you need it.

HAND TOOL SELECTION

Buy good-quality tools. You probably don't need the same range of tools a mechanic does but at least consider a similar line. A little less finish and a chrome-vanadium alloy steel with a forged design are the best bet. It takes a number of years and dollars to build up a suitable tool selection, but with the cost spread over a number of years, it becomes easier to handle. We all want the right tool at the right time for the right job. Sometimes this is not possible at the start. Don't buy specialized tools at the beginning. Start with a good, basic repair kit.

A set of combination wrenches that have a box end and an open end are a good place to start. Two adjustable wrenches (a 6 or 8 inch and a 10 or 12 inch) will make good companions for the combination wrenches so that you can remove nuts from bolts.

Screwdrivers are next. Look around at what you already have. A good assortment is possible with the replaceable bit set that is stored in the handle. The most common type ends are the flat and the Phillips. Buy one heavy-duty screwdriver so that you can use it as a pry bar.

For pliers, the combination, adjustable jaw and vice-grips (which you probably already have) will do.

A ball-peen hammer, a 16-ounce hammer, a 1-inch cold chisel, and a tapered punch will often be of help. A hacksaw is a valuable cutting tool. Buy one with an adjustable handle so different blade lengths can be used. Use a flat file to remove burrs and flashings.

Buy one spark-plug socket of the size to fit and a style that a box end wrench can be used to turn it.

For measuring tools, use a steel rule or push pull tape. A flat blade feeler gauge and a spark plug wire type of gauge will do. A good-quality tire pressure gauge will help.

You will be able to save money by doing your own oil changes and grease jobs, but you need some special equipment. The most important item is the oil filter wrench. A grease gun is next; buy the one that takes the grease cartridge. You will also need a set of car stands of suitable strength to support the car, a drain pan for the used oil and a combination funnel, and a can opener. Check the local regulations for disposing of used oil. As you become more proficient in tool use, you might want to add a ratchet wrench set with suitable drivers and sockets. These are bought with a box so it help to keep the pieces together. Expand the punch line to include a center punch, drift punch, and a set of pin punches. Study the tool catalogs and let your family members know that you appreciate tools. You know what you need so tell them.

For portable equipment a 3/8 variable speed drill is invaluable, for drilling, buffing, grinding, and polishing. Buy high-speed drill bits of the size you need and not the drill size assortment.

CARE OF TOOLS AND EQUIPMENT

You probably practice good housekeeping at work so continue it into your personal work place. Store your tools so they can be easily located. A pegboard is a good way to group them. A tool box is great if you have the money. Construct a workbench and, if space is at a premium, make the bench portable or make it so that you can fold it against the wall.

A vise is a terrific holding tool so mount it solidly even on the wall. Storage space for oils, filters and parts can be located on the walls nearer the ceiling. This is your place so make it pleasant and comfortable to work in.

Wipe dirty tools clean before you put them away. A slightly oily cloth will do the job and help prevent the tools from rusting. Use tools that are in good condition. Striking tools such as punches and chisels can mushroom out at the ends after a period of use. Dress these ends on a grinder or a file to eliminate the possibility of tool chips breaking and flying.

Check your hammer handle to see that it is tight in the head. See that the striking surface is smooth and strike the work with the full, flat face. Glancing blows can chip the hammer face.

Wrenches must be the exact size for the bolt to prevent slippage. Six-point wrenches might help on rounded-off bolts. Pull on a wrench; don't push. If you must push, use the base of your palm to prevent "barking" your knuckles. When using an adjustable wrench, snug it up tight and apply the pulling force to the stationary jaw side of the wrench. Which wrench to use is something best learned by experience, but you will not find it difficult. Tool catalogs often have good information regarding tool use.

Screwdrivers are basically intended for tightening or loosening screws and if you do use them as pry bars you will break the tips and bend the shanks. To service a flat tip, square the tip and grind the sides almost parallel. Keep the tip cool when grinding or else the heat will draw the temper from the tip and it will become soft. The tip should fit the slot snugly and not be any wider than the screw. If the sides taper toward the end, the screwdriver will raise out of the slot when the twisting force is applied. A Phillips tip can be serviced by maintaining the same tip angle but the tip must fit properly in the screw slots.

The file and the hacksaw are probably the most abused of the cutting tools. Never use a file without a handle. You might drive the tang of the file into your hand. Files are available in single or double cut, and in a variety of shapes to suit the job being done. Clamp the work in the vise with the top of the work level with your elbow. If the part has a finished surface, use a few layers of masking tape to prevent marking. When taking a heavy cut or rough filing, use a stroke of about 50 degrees and use enough pressure to keep the file cutting. For finish filing, use a fine-cut file and light pressure. Release the pressure on the back stroke and raise the file to prevent dulling the cutting edge of the teeth.

Draw filing is a method used to produce a smooth finish on the work. A single cut file is drawn crosswise on the work, with pressure applied on both forward and backward strokes. Keep the file clean to prevent small particles from scratching the work. A narrow block of wood pushed across the file, parallel to the teeth, will clean them. Rubbing chalk into the file teeth might prevent scratching. Keep the file clean of cuttings. Store files in a rack to protect their cutting edges.

When using a hacksaw, place the blades in so that the teeth point away from the handle. This means that the cutting action is on the forward stroke. Use sufficient pressure so that the teeth will cut, and use the full length of the blade at about 40 strokes per minute. A blade with 18 teeth per inch is good for all-purpose work. Two or more teeth should be in contact with the work at all times.

Pliers are not to be used for tightening or loosening nuts or bolts. They also are not to be used for or on brass fittings. Don't use them on hardened surfaces because it dulls the teeth and the jaws lose their grip. Use pliers only when no other tool will do the job.

In addition to oils and greases, you will need some other supplies: wiping rags or paper towels to clean up spills; a metal, airtight container for disposing of oily rags; air, oil, and gas filters; electrical and masking tape; cleaning solvent; a spray bottle you can fill with rust penetrating oil; assorted nuts, bolts, washers (both flat and lock), and cotter pins.

There are lots of new gasket sealers on the market. Check on chemical welding supplies such as epoxy. A soldering gun and solder for making up good electrical joints and a bench grinder with a buffing wheel for cleaning parts are real timesavers.

If you are going to buy test equipment, make sure it will work on your car. You should consider a torque wrench if you want to get into the real take-apart-and-reassemble area of service work. Basically, you are only limited by what you want to do and the tools you have for the job. Get in to this slowly and enjoy the jobs you can do. If you remove the faulty part and replace it with a rebuilt unit that you have bought as an exchange, that is fine. You've got the car running, and that's what it's all about.

SAFETY

You must practice personal and property safety in your work area. The service and maintenance of the automobile presents conditions that can be a safety hazard to life and limb. An accident can turn a pleasant experience into tragedy. The possibility of a fire is one of the greatest dangers. A shop without a working foam or carbon-dioxide extinguisher is not a safe place. Make an extinguisher your first purchase, and know how to use it.

Store gasoline only in approved containers. You already have a gas tank in the car. How much more do you need? Label any other containers that hold oil, antifreeze, solvents, etc. Do not use gasoline to wash parts or tools; use solvent. Dispose of oily rags.

Do not operate a car engine indoors. Have a flexible exhaust tube to carry the fumes outside. Carbon monoxide is a killer. This gas is odorless, colorless, and tasteless, and it is a product of combustion. Don't operate the car engine in a closed garage. In three minutes, a small garage can become filled with enough carbon monoxide to kill you.

Take care when removing the radiator cap from a hot or boiling radiator. The hot coolant can shoot up directly at you and painful burns can result. If you cannot comfortably keep your hand on the top of the radiator, leave the cap on until the radiator cools down. Be careful around the engine when the cap until the radiator fan is turning. Don't get your fingers in the way of the blades. Many times you will have to make adjustments with the engine running. Watch out for the hot exhaust system. At the same time, watch out for the fan belts because they can pull in a piece of loose clothing.

Battery acid is corrosive. Use rubber-backed carpet remnants as fender covers when servicing the battery. Neutralize skin or clothing with baking soda and water. Do not smoke around the battery because an explosion could result. Hydrogen gas is generated as a natural result of battery action, and it is possible to ignite this gas with a spark or flame. Get in the habit of disconnecting the battery ground strap when working under the hood. Don't spark a battery with a screwdriver or a pair of pliers to see if it has any charge. You might end up with no battery at all.

Don't trust any hydraulic or mechanical jack support an automobile, and never get under a car when it is only on a jack. Use car stands of the rated weight load to support the car. When two wheels are off the floor, block the other wheels to prevent the vehicle from rolling.

Have a first-aid kit handy so that you can look after burns and cuts immediately. Many parts of the engine are hot so learn to avoid personal contract with them. Keep your hands away from the fan; it cuts meat readily. Do as much work as you can on the engine while it is not running.

When grinding or buffing, use a face shield or safety glasses. They haven't yet made a glass eye that you can see through. Make sure that electrical equipment is properly grounded. Don't break the ground prong off or you will complete the circuit. Check all cords for chaffing or breaks and have them repaired. Service work can be very satisfying, but it can also become a hazard if procedures are not followed in a responsible manner. There is absolutely no place for poor workmanship. You endanger your life and the lives of others with slipshod work. If you are not willing to do the best work on your own vehicle, then leave it to the commercial shops before you become a statistic.

Be safety alert all the time and consider that accidents can happen to you. They don't just happen to someone else. With just a bit of planning and foresight, you can enjoy service work. Remember the ABCs: Always Be Careful.

Make up an emergency kit for the road. See that you have some basic tools, spare headlight and bulbs, booster cables, and fan belt. A fire extinguisher and a set of flares is a good investment.

IDENTIFYING YOUR CAR

You must present detailed information when ordering parts. It is not enough to know the year, model, and type of engine. See Figs. 1-1 through 1-9 and Table 1-1. Locate the indentification plate that lists the year, model, serial number, engine number, and paint code. Additional unit serial numbers will help you to identify component parts. You can buy parts from the automotive dealership, jobbers, or the local wrecking yard. If you cannot buy the parts to rebuild a certain component, you can buy an exchange unit instead. This is a good way to get started. As you gain more experience, you can do more of your own work.

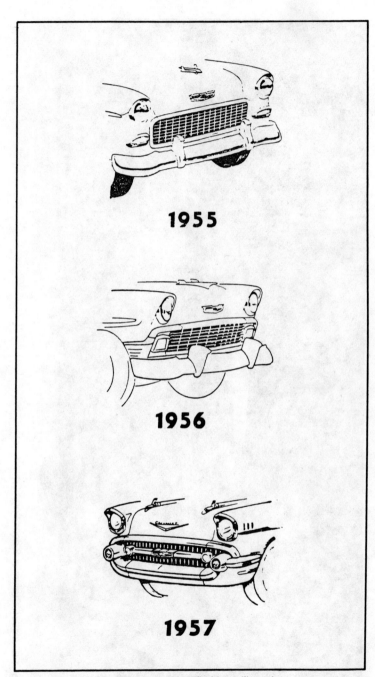

1955

1956

1957

Fig. 1-1. Classic Chevies can be identified by grille style.

Fig. 1-2. A 1955 Chevrolet 2-door Sport Coupe.

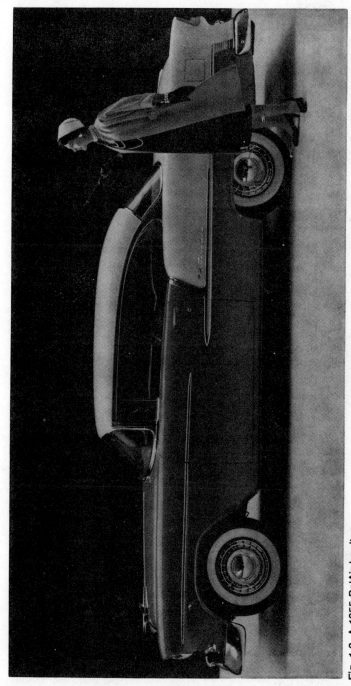

Fig. 1-3. A 1955 BelAir hardtop.

9

Fig. 1-4. A 1956 Chevrolet 4-door BelAir hardtop.

Fig. 1-5. A 1956 Chevrolet 4-door sedan.

Fig. 1-6. A Chevrolet 3100.

Fig. 1-7. A Chevrolet 3200.

13

Fig. 1-8. A 1957 Chevrolet.

14

Fig. 1-9. A 1957 Chevrolet 2-door BelAir hardtop.

15

Table 1-1. Model Numbers and Engine Identification Numbers.

1955 Model No.	Body Style	Model
551000	(2400, All	BelAir
	2100)	Two-Ten
551200	(1500) All	One-Fifty
551011D	(2402) Sedan—2-Door	Bel Air
551019D	(2403) Sedan—4-Door	Bel Air
551037D	(2454) Sport Coupe	Bel Air
551067D	(2434) Convertible	Bel Air
551062DF	(2409) Station Wagon "Beauville" (4-Door)	Bel Air
551064DF	(2429) Station Wagon "Nomad" (2-Door)	Bel Air
551011	(2102) Sedan—2-Door	Two-Ten
551011A	(2124) Club Coupe	Two-Ten
551019	(2103) Sedan—4-Door	Two-Ten
551037	(2154) Sport Coupe	Two-Ten
551062F	(2109) Station Wagon "Townsman" (4-Door)	Two-Ten
551063F	(2129) Station Wagon "Handyman" (2-Door)	Two-Ten
551211	(1502) Sedan—2-Door	One-Fifty
551211B	(1512) Utility Sedan	One-Fifty
551219	(1503) Sedan—4-Door	One-Fifty
551263F	(1529) Station Wagon "Handyman" (2-Door)	One-Fifty
551271	(1508) Sedan Delivery	One-Fifty

1956 Model No.	Body Style	Model
561000	(2400, All	Bel Air
	2100)	Two-Ten
561200	(1500) All	One-Fifty
561011D	(2402) Sedan—2-Door	Bel Air
561019D	(2403) Sedan—4-Door	Bel Air
561037D	(2454) Sport Coupe	Bel Air
561039D	(2413) Sport Sedan	Bel Air
561067D	(2434) Convertible	Bel Air
561062DF	(2419) Station Wagon— "Beauville" (9 Pass.)	Bel Air
561064DF	(2429) Station Wagon— "Nomad" (2-Door)	Bel Air
561011	(2102) Sedan—2-Door	Two-Ten
561011A	(2124) Club Coupe	Two-Ten
561019	(2103) Sedan—4-Door	Two-Ten
561037	(2154) Sport Coupe	Two-Ten
561039	(2113) Sport Sedan	Two-Ten
561062F	(2109) Station Wagon— "Townsman" (4-Door)	Two-Ten
561062FC	(2119) Station Wagon— "Beauville" (9 Pass.)	Two-Ten
561063F	(2129) Station Wagon— "Handyman" (2-Door)	Two-Ten

Model No.	Body Style	Model
561211	(1502) Sedan—2-Door	One-Fifty
561211B	(1512) Utility Sedan	One-Fifty
561219	(1503) Sedan—4-Door	One-Fifty
561263F	(1529) Station Wagon— "Handyman" (2-Door)	One-Fifty
561271	(1508) Sedan Delivery	One-Fifty
	(2934) Convertible (2 Pass.)	Corvette

1957

Model No.	Body Style	Model
571000	(2400, All	Bel Air
	2100)	Two-Ten
571200	(1500) All	One-Fifty
571011D	(2402) Sedan—2 Door	Bel Air
571019D	(2403) Sedan—4 Door	Bel Air
571037D	(2454) Sport Coupe	Bel Air
571039D	(2413) Sport Sedan	Bel Air
571067D	(2434) Convertible	Bel Air
571062DF	(2419) Station Wagon "Beauville" (4 Pass.)	Bel Air
571064DF	(2429) Station Wagon "Nomad" (2-Door)	Bel Air
571011	(2102) Sedan—2 Door	Two-Ten
571011A	(2124) Club Coupe	Two-Ten
571019	(2103) Sedan—4 Door	Two-Ten
571037	(2154) Sport Coupe	Two-Ten
571039	(2113) Sport Sedan	Two-Ten
571062F	(2109) Station Wagon "Townsman" (4-Door)	Two-Ten
571062FC	(2119) Station Wagon "Beauville" (9-Pass.)	Two-Ten
571063F	(2129) Station Wagon "Handyman" (2-Door)	Two-Ten
571211	(1502) Sedan—2-Door	One-Fifty
571211B	(1512) Utility Sedan	One-Fifty
571219	(1503) Sedan—4-Door	One-Fifty
571263F	(1529) Station Wagon "Handyman" (2-Door)	One-Fifty
571271	(1508) Sedan Delivery	One-Fifty
	(2934) Convertible (2-Pass.)	Corvette

Year	Engine	Suffix Letter [1]	
1957	6-235 [2]	A	[1] —Engine identified by first letter after engine number.
	6-235 [3]	B	[2] —Manual shift transmission.
	V8-265 [2]	C	[3] —Powerglide.
	V8-283 [2]	E	[4] —Turboglide.
	V8-283 [3]	F	[5] —Manual shift transmission and heavy duty clutch.
	V8-283 [4]	G	

MODEL YEAR IDENTIFICATION

The year, model and serial number, engine number, trim, and paint numbers are stamped on a metal plate located on the top of the cowl on the right-hand side under the hood or on the left-front door pillar.

The serial number identifies the vehicle, and the engine number identifies the engine. It is quite possible that the serial number and the engine number will not match for the model style because someone has exchanged parts to keep the car running. If you want to go back to square one, be prepared for a lot of work. Sometimes it's better to run what you have and look for the parts rather than have a lifetime project on your hands.

Canadian serial numbers indicate the year of the car by the first digit: 5 for 1955; 6 for 1956; 7 for 1957. American models are the first two digits following either one or two letters. Thus 55 for 1955; 56 for 1956; and 57 for 1957. When the first letter is preceded by a V, it indicates a V-8 engine. The second letter indicates the model: A for series 1500 Special; B for series 2100 Deluxe; C for series 2400 Bel Air.

ENGINE NUMBERS

Engine numbers are located on the right side of the block at the rear of the distributor on the six-cylinder engine and on the front right side of the block on the V-8 engine. Present these numbers to a knowledgeable partsman and you should be able to deduce whether the engine, transmission, and rear axle are matched.

Automatic transmissions used the engines with hydraulic lifters. The rear axle serial number is located on the front side of the differential carrier.

Other serial numbers that might help for authentic restoration are located on the generator body just forward of the armature terminal and the starting motor body below the solenoid.

Chapter 2

Preventive Maintenance

Preventive maintenance is regular service based on the amount of miles you drive. This means that you have to keep a record of the mileage and the service you perform. Basically, you have to start a service diary. The owners manual or shop service manual will provide a lubrication chart that shows the mileage at which various parts should be lubricated and the type of lubricant to use. See Fig. 2-1. Use a notebook to list what you have done and at what mileage.

The owners manual will also suggest when the tires should be rotated, the battery water level checked, the fluid levels in the transmission and differential checked, and the time for an engine tune-up. Don't wait until you have a problem with the car; get in the habit of immediately checking little tip-off signals like noises and squeaks. A poorly maintained car will cost you $$$. Not only does such a car burn more gas but a simple problem can turn into an expensive repair. Preventive maintenance makes break downs less likely and allows for safe and economical driving.

INSPECTIONS

Get in the habit of doing a walk-around before you get into your car. Check for broken lamps, low tire pressure, or wet spots under the engine area that could indicate a burst hose. Start the engine. Warning lights should come on when starting and then go out. Try the brake pedal. It should be firm when stepped on. There should not be any unusual noise with the engine running or when the car starts moving.

Once a month make a more thorough inspection. Move the car up one car length and check for wet spots. You could place a sheet of paper under the car and on the garage floor for overnight if you want clear, color definition. A dark brown or black stain under the engine indicates an engine oil leak. Reddish or pink stains in the area of the transmission indicate an automatic transmission leak. A dark brown stain under a standard transmission is a similar problem. Stains under the differential housing could be caused by a leaking seal.

Radiator coolant will show a spot the same color as the antifreeze that is being lost. Power steering is the same color as the automatic transmission because it uses the same fluid. A brake fluid leak is whitish and sometimes can be seen at the wheels or at the master cylinder. Most of these fluid leaks are caused by seals that are not holding the necessary fluid.

Raise the hood and check the appropriate dipsticks. The brake fluid level is checked by removing the cap. The brake fluid should be about 3/8 of an inch below the top. On the standard transmission and the differential, you will have to get under the car and remove the plugs to check the fluid levels. Check the coolant level.

If the leaks are severe and not just a wetness, the necessary seals or hoses should be replaced. Transmission and differential leaks can be caused by wear on the connecting yokes. A leaking transmission oil pan will need a gasket replacement. A leaking master cylinder or wheel cylinders must be repaired immediately.

Check the tires for cuts and bruises. If the center tread is wearing faster than the edges, check for overinflation. If the edges wear faster, it might be due to underinflation or hard cornering. Misalignment will cause one edge of the tread to wear faster than the other. Tires that are worn to the tread wear indicator across two or more grooves should be replaced. Check the spare tire pressure.

Clean the lights and try the light switch on all positions. Try the dimmer switch and check the head lamps. Work the signal lights. Replace any bulbs that are not shining. Check the windshield wipers for cracks and replace as necessary.

Check the level of the battery water and add distilled water if the water in your area has a high mineral content. If the battery cables are corroded, they should be cleaned with a baking-soda-and-water solution. Rinse with clear water. Check the drive belts

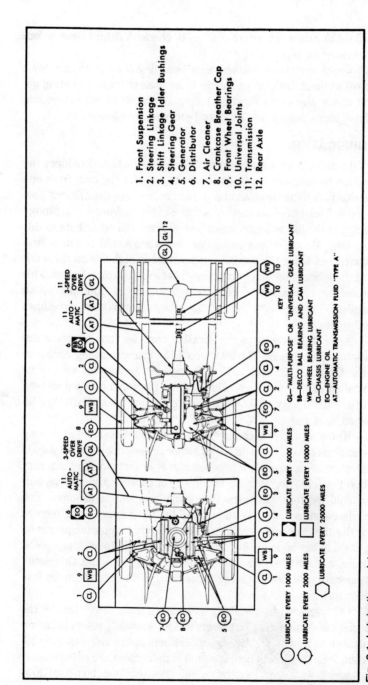

1. Front Suspension
2. Steering Linkage
3. Shift Linkage Idler Bushings
4. Steering Gear
5. Generator
6. Distributor
7. Air Cleaner
8. Crankcase Breather Cap
9. Front Wheel Bearings
10. Universal Joints
11. Transmission
12. Rear Axle

KEY

GL—"MULTI-PURPOSE" OR "UNIVERSAL" GEAR LUBRICANT
BB—DELCO BALL BEARING AND CAM LUBRICANT
WB—WHEEL BEARING LUBRICANT
CL—CHASSIS LUBRICANT
EO—ENGINE OIL
AT—AUTOMATIC TRANSMISSION FLUID "TYPE A"

◯ LUBRICATE EVERY 1000 MILES
◯ LUBRICATE EVERY 2000 MILES
⬡ LUBRICATE EVERY 5000 MILES
▢ LUBRICATE EVERY 10000 MILES
⬡ LUBRICATE EVERY 25000 MILES

Fig. 2-1. Lubrication points.

21

and replace them if they are cracked or checked. Adjust them before they start to slip

Check the shock absorbers by bouncing the car up and down at each wheel. If the car bounces more than twice after letting go, the shock absorbers need replacing. Worn shock absorbers can cause unnecessary tire wear and steering problems.

LUBRICATION

In the engine, oil separates the moving parts and reduces the friction between them. It also helps to transfer the heat from one engine part to another and on to the coolant. An insufficient flow of oil will lead to rapid wear or seizure of the moving parts. Sludge or dirt in the oil passages and a low oil level will contribute to this problem. Most engine wear occurs not when a cold engine is first started but when the engine temperature is high the oil thins out and the lubricating film breaks down between the metal parts. This causes an expensive problem that is known as "burning up an engine." The bearings, connecting rods, and crankshaft sometimes end up as one.

An oil filter removes the abrasive particles that can cause excessive engine wear. The filter should be changed at regular intervals. On the 1956 and 1957 V-8 engines, an adapter kit can be installed so that the disposable spin-on filter can be used instead of the cartridge type. The six-cylinder engine uses a cartridge filter as does the 1955 V-8.

Knowing when to change oil is a difficult decision, but driving conditions have a lot to do with change periods. Color is not a good indicator because some oils become gray or dirty looking in a very short time. Driving in dusty conditions or short-trip operations during cold weather will necessitate more frequent oil changes. You might consider changing the oil and the filter every three months for the above conditions. For standard driving, you can operate on a 6000-mile, oil-and-filter change period. Oil lubricates, cools, cleans, and seals. Maintain a regular oil change schedule. Oil doesn't wear out but it does become contaminated and the additives lose their qualities.

Oil is graded by viscosity on a rating scale developed by the Society of Automotive Engineers (SAE). Viscosity refers to the resistance to flow, with the higher numbers indicating a thicker oil. A single oil that meets both low- and high-temperature requirements is known as a multi-viscosity oil. Oils that meet special low-temperature requirements have the letter W following the viscosity rating. Common viscosity ratings are SAE 10W, SAE 20W, SAE 30 and SAE10W30.

The American Petroleum Institute (API) classifies oils under service classification, and these indicate the severity of service. For gasoline engines, these are SA, SB, SC, SD, SE, and SF. The SF rating provides maximum protection against rust, corrosion, wear, oil oxidation, and high-temperature deposits that can cause oil thickening. You will find the viscosity rating and service classification stamped into the can lid. Buy the best oil you can and leave the lubrication planning to the oil chemists.

Synthetic oils have been around in the aircraft industry since World War II, although they are just being introduced for automotive use. They are chemically formulated and usually contain friction-reducing additives. In addition, they have a very low pouring temperature and a very high breakdown temperature. Because such oils don't break down easily, the formation of sludge is reduced and the oils stay cleaner longer. Manufacturers refer to synthetic oils as extended-life oils because the oil-change period can be extended to 25000-mile intervals. There is also a reported advantage of increased gas mileage. Due to limited use, the cost of this oil is about four times the cost of regular oil. You will have to decide if your driving habits and the amount of miles you drive warrant the use of this oil.

If you decide to change to synthetic oil, you must use a compatible oil filter. Regular oil filters are not suitable. Manufacturers are introducing high-performance oils so its a case of cost over mileage. Table 2-1 can be used to determine the correct viscosity of oil to use. Except in cases of extreme low temperatures, an SAE 10W-40 oil with an American Petroleum Institute classification of SE will serve under all operating conditions.

To change the oil and the filter, you will need a wrench to fit the oil-pan drain plug and a filter wrench if the engine has been changed to a spin-on filter. A plastic dish pan makes a good container for draining the oil. See what the regulations in your area say about disposing of old oil. You will also need a funnel, a can opener, some clean rags, and a piece of foam-backed carpet to protect the fender.

The engine should be at operating temperature so that the dirt and foreign material in the oil will drain out better. If the oil filter is located topside, you might be able to slide the pan under with-

Table 2-1. Oil Viscosity.

Lowest Temperature Anticipated	Recommended Single Viscosity	Recommended Multi-Viscosity
Above − 10°F (− 23°C)	SAE 30	SAE 10W-30
	or	20W-40
	SAE 40	10W-40
Below − 10°F (− 23°C)	SAE 20W-20	SAE 10W-30
		10W-40
	SAE 10W	SAE 5W-20
	SAE 5W	5W-30

out raising the vehicle. If the filter is underneath, it helps to raise the front of the vehicle.

Remove the drain plug and let the oil drain into the dish pan. Wipe the drain plug clean, and see that the gasket is in place and not cracked. Replace it if necessary. Remove the filter if it is on the bottom side. It has hot oil in it so be careful. If the filter is in a filter case, make sure the old gasket comes off. Replace the filter and the gasket. Replace the case and tighten the hold-down bolt. With the spin-on filter, lubricate the gasket and turn the filter half a turn after it contacts the block. Do not use the filter wrench; tighten by hand. Install the drain plug.

Remove the oil filler cap and fill the crankcase with the correct type and amount of oil. The filter takes an additional quart. Service the topside oil filter if so equipped. Clean the case properly before putting in the new element and use a new gasket. Start the engine and let it idle. The oil indicator light must go out after a few seconds. Check for leaks at the drain plug and at the oil filter. Check the oil level on the dipstick; it should be between Full and Add.

The purpose of greasing the chassis is similar to that of using oil in the engine (namely to separate the metal parts and prevent them from running dry). How often you lubricate the chassis will depend on the weather conditions and the amount of miles you drive. Rain, mud, and water tend to drive the lubricant out of the joints and also contaminate the remaining grease. You might decide to lubricate the chassis every time the oil and filter are changed or follow a spring and fall schedule. This is also a good time to inspect some of the other running gear.

You will need a grease gun with a flexible end and a cartridge of moly lithium EP grease or synthetic grease. The grease will last a long time so buy the best. Study the lubrication diagram (Fig. 2-1) and you will find the grease fittings sooner. You might find it easier to place the vehicle on stands.

Wipe the grease fitting clean before attaching the grease-gun hose. Some joints will have a rubber seal around them so add just enough lubricant to swell the seal. Don't burst the seal because contaminants will enter. On joints without a seal, pump the old grease out and wipe away the excess. Replace any grease fittings that don't take the grease in or will not hold the grease.

Examine the front end parts for signs of wear and road damage. A bent tie rod is a sure sign that steering problems and tire wear are not far behind. Tend to these problems immediately. Grease the clutch compensating shaft. If the universal joints are of the sealed type, they must be disassembled, cleaned, and lubricated every 25,000 miles. If they have lubrication fittings, give them just a touch of grease. Don't break the seals with over lubrication.

Check the other chassis points that use rubber bushings. If they are beginning to squeak, don't lubricate with oil; use a silicone spray. Rear spring leaf liners that are worn through should be replaced with new material. Spraying doesn't help to stop a metal to metal squeak. Use 30-weight engine oil to lubricate any other pivot points or give them a light smear of grease.

The steering gear oil level should be checked. If there are no signs of external leakage it should be okay. Do not over lubricate because excess lubricant will be forced up to the steering wheel area. Top up with a multipurpose gear lubricant. Find out why this unit is low on lubricant and correct the problem. On models with power steering, add automatic transmission fluid "type A" to bring the level up to the full mark or just below. Do not overfill. Again check where the oil is going if you have to keep topping up this reservoir.

The lubricant level in the differential and the standard transmission should be checked if external signs of leakage are present. If the lubricant is cold, it will probably be below the oil filler plug hole by about 1/2 inch. Leaks should be repaired. These parts are two expensive to burn out due to lack of lubrication. Drain these units once a year and refill with clean lubricant. Check what comes out with the oil. If it's a part of a tooth, keep in mind that the tooth fairy doesn't operate in this hemisphere.

On the 1957 models with a standard transmission, smear the gearshift idler bushing with a bit of grease to facilitate shifting of second and high gear.

Check the levels of the fluid in the automatic transmissions with the temperature up to operating level and the selector in Neutral, parking brake set and engine idling. Wipe the area around the filler tube and the dipstick. Remove the dipstick, wipe dry, and then reinstall the dipstick. Remove and read fluid level. Add only if below the Full mark. Use automatic transmission fluid "Type A." To extend the life of both the Powerglide and the Turboglide, drain and refill the transmission fluid every 25,000 miles.

If the vehicle is on stands, this is a good time to check the wheel bearings for endplay and noise. These bearings should be cleaned and repacked about every 25,000 miles. We have better greases on the market than we had in the '50s and we have been able to extend the lubrication periods. To lubricate the front wheel bearings, it is necessary to remove the wheel and hub assembly.

Pry off the grease cap and remove the cotter pin, the adjusting nut, and the thrust washer. Pull the hub off; if the brake shoes stick in the drum, release the shoes. This is probably a case of badly scored drums so you might end up doing some brake work. Drive out the inner bearing and grease seal. A hammer handle will do the job, but in time it gets the hammer handle marked up.

Clean everything thoroughly. Don't spin the bearings with air pressure. Examine the bearing cups for discoloration, pitting, scoring, cracks, and chipping. Examine the ball bearings for similar defects. The cage should not be bent. If the bearing or cone needs replacing, you must replace as a set. Replace the seals. Grease the bearings by forcing the grease (lithium-base) between the bearing ball rollers. Have some grease in the cup of your hand and scrape the bearing full. Install the inner cone and a new grease seal. Wipe the spindle clean and lightly grease it. Do not pack the hub between the inner and outer bearings because this grease will work out into the brake drums and brake shoes. Do any necessary brake work, and fix leaking cylinders, worn shoes, or scored drums.

Install the hub and drum assembly. Install the outer bearing and thrust washer and adjusting nut. Tighten the adjusting nut while turning the drum until all the end play is removed. Back off 1/6 turn. Check for end play; there must not be any. Insert the new cotter pin in either the vertical or horizontal spindle hole.

Cut off any extra length so the pin does not interfere with the dust cap. Lightly grease the inside of the dust cap and put it on. Replace the wheel and the wheel cover. Let the car down.

Let's do some under-the-hood checks. Service the air cleaner and use the same oil as for the engine. The oil filler cap should be washed in solvent and tapped on a flat surface to remove any dirt. Lubricate the mesh with engine oil. Use a drop of oil on the generator, distributor, and starter solenoid linkage. The water pump is of the sealed-bearing design and cannot be lubricated.

Check the manifold heat-control valve for free movement. It should turn freely. Use a carbon-dissolving spray to loosen it. Do not use an oil base as you compound the problem. Tapping with a small hammer will help free the shaft in its bearings. Then soak it up well with carbon cleaner.

When you are under the hood, get in the habit of checking other parts for wear or leaks. Give the radiator hoses a squeeze. If they feel soft, they will soon need replacement (better at home than on the road). The cost of the lost antifreeze will pay for hose; won't it? Check the belt(s). Increase tension if necessary. Check the battery. The top should be clean and the cables should be tight. What about the level of the coolant in the radiator? Preventive maintenance fixes minor problems before they turn into major expenses.

Don't over lubricate. Lubricate only where squeaks develop or where conditions indicate that the addition of lubricant is desirable for easier operation. Remove the old, contaminated lubricant and apply a light coating of the new. There are new products on the market but you can use engine oil on rotating parts and hinge pins. Parts that might touch clothing should have a nonstaining, waterproof white grease. Lubriplate works on hinges and latches on the doors, hood and trunk.

Apply the new lubricant and work the parts to distribute it, and then wipe away the excess. Lubricate the seat tracks and mechanisms as well as the hinges on the sun visors, instrument compartment door, door holds, and torque rod ends. Apply oil sparingly to the convertible top mechanism and have a cloth ready to catch the drips. Spray all the rubber with silicone spray or rubber preservative. You can also use this spray on the locks. Squirt it into the lock and work the key around to distribute the lubricant. The window regulators should be lubricated with Lubriplate, and at the same time you can do the door lock parts. This means removing the upholstery panels, but its possible that they have not been lubricated since factory assembly. The windshield wiper motor and transmission spools should also be done. If you have a look at these body parts about twice a year, things will run much smoother, quieter, and longer.

EXTERIOR AND INTERIOR CAR CARE

You will greatly increase the life of your vehicle if you wash and wax it regularly. In areas where salt is used on the streets and highways, a rust inhibitor should be sprayed on the underside when the vehicle has been completely washed. There isn't much sense in having a good chassis if the body is rusting off. If the vehicle is stored outside, the only way to give it any protection at all is to preserve the finish. Chemicals in the air tend to oxidize the paint finish giving it a chalky look. After washing, a coat of wax applied on the clean surface will help preserve the finish.

The automatic car wash doesn't do a particularly good job but it's a start. Wash at home with a good spray nozzle or go to a commercial wash and use their equipment. If you do this twice a year, you should be able to maintain the cleaning at home. Don't wash in the direct sunlight or the finish will streak. Spray the underside clean. This includes the fender wells, back sides of bumpers, and the underside of the engine, power train, and suspension.

Wash the inside of the engine compartment. The engine itself should be clean from previous cleaning so all it needs is a once-over. Don't short out the electrical system. A clean engine is a better-looking and a cooler running unit.

Now to the body wash. Use a car soap or a wash-wax. If a wash-wax is used regularly on an originally clean finish, you might not have to wax the surface. Follow the product instructions. Flush all dirt and dust loose from the body. Make up the wash solution in a suitable container. Wash one section at a time with a car brush or sponge. Rinse. Difficult spots of tree or bird droppings and tar should be removed with a commercial cleaner. When the entire car is washed, rinse it down and dry it with a chamois (buy a good one or you will end up buying a number of cheap ones). Clean the glass inside and out. Vacuum the interior. Vinyl cleaner can be used to protect the finish.

To determine whether the finish needs waxing, remember if the water beaded on the hood during the washing (just like the commercial). If the finish is oxidized, you will have to apply a liquid cleaner before waxing. In extreme cases, a paste compound or rubbing compound will have to be used. Yes, that's the original color under all that film. Now comes the part where you need the elbow grease. Apply the wax one section at a time and polish it out using a circular motion. Overlap the sections to blend everything together.

Before you are tempted to use an electric drill and a buffing or polishing pad, don't or you will go through the finish.

To clean the upholstery, use warm water and liquid soap. Rub gently and rinse with clear water. Wipe dry. Soiled spots can be removed with good-quality fabric cleaner. The carpeting can be looked after the same way as your living room rug. Shampoo about once a year with household carpet cleaner. If you are using floor mats to protect the carpet, let the carpet dry first. If it doesn't dry, mildew will set in.

TIRE ROTATION

The tires used on the 1955-56 models are tubeless, 6.70-15 in. size and have a 4-ply rating. In 1957, a "low pressure" tire of the 7.50-14 size and with a 4-ply rating was introduced. Correct tire pressure for the 15-inch size is 24 pounds, front and rear, and for the 14-inch size it is 22 pounds front and rear. Tire sizing and construction has changed a considerable amount, and there are better tires on the road than we had in the '50s. You should not mix types of tires as this will affect handling and control.

There are three types of tires: bias ply, bias belted, and radial. If you are a low-mileage driver, you can probably get by with the bias-ply tire. If possible, replace tires in pairs. If you have to replace one tire only, pair it up with the tire having the most tread. When installing new tires, have them balanced; they will last longer and the vehicle will handle better.

Normal tire wear is uneven between the front and the rear tires. The front tires steer the car and the rear tires drive them. An uneven weight distribution can also cause the tires to wear unevenly. To equalize wear, the tires should be rotated on or about the 5000-mile mark. By placing the spare tire in rotation with the other four, you get about 20 percent more total tire mileage before replacement tires must be purchased. Tire rotation order depends on the type of tires and the amount of wear. The following order is for bias ply tires.

Move the left, front tire to the left rear, left rear to right front, right front to spare, spare to right rear, and right rear to left front. Use the jack and wheel wrench out of the trunk if you have no other equipment. Always be careful positioning the jack. Block the wheels so the vehicle will not move. If you want to rotate tires in order of wear, use this procedure. Put the tire with the best tread on the right rear, second-best tire on the left rear, third-best tire on the right front, and the poorest on the left front.

Tighten the lug nuts securely before you put the hub caps or wheel discs back on. Wheel rotation will not correct tire wear that is due to misalignment or unbalance. Keep tire pressure at the proper inflation to prevent shoulder or center-tread wear.

Chapter 3

Engine Service

The six-cylinder in-line engines used in 1955 were available in two different models. The engine used in standard transmission and overdrive models was equipped with solid-valve lifters. The Powerglide models were equipped with hydraulic valve lifters. With different camshafts, the standard model developed 123 horsepower while the Powerglide-equipped models developed 136 horsepower. The engines have a displacement of 235.5 cubic inches, a 3 9/16-inch bore, 3 15/16-inch stroke and 7.5:1 compression ratio.

By 1957, only one six-cylinder engine was provided for all passenger models and types of transmissions. The Blue Flame Engine is similar to the above except the compression ratio was changed to 8.00:1 (producing 140 horsepower). The high lift cam was used with hydraulic valve lifters. A full-pressure lubrication system is used on all models. A gear-driven pump maintains 35 psi pressure lubrication. Full pressure is provided to the main and connecting rod bearings, the camshaft bearings, and the valve mechanism.

The early V-8 engines also had the different valve lifters (depending on transmission models), but by 1957 they were all equipped with hydraulic lifters. The displacement for one of these was 265 cubic inches and for the other it was 283 cubic inches. The standard or overdrive transmission equipped V-8 engines used the former while the automatic transmission models used the latter. An optional Super Turbo Fire 283 with four barrel carburetor and dual exhaust was also made available.

The base engine developed 162 horsepower, with a compression ratio of 8.00:1, and weighed less than the Blue Flame 6. The 3 3/4-inch cylinder and 3-inch stroke was known as the "over-square" design. This was the 1955 design and it has paved the way for every short-block Chevrolet.

COMPRESSION

You should take a compression test of the engine when you do a tune-up. If you are planning to buy a car, it's important that you find out what shape the engine is in, and this is the best way to do it. Forget what the present owner has to tell you (even if he has bills to prove it). Without taking an internal engine check, you are asking for a lot of problems.

The equipment you will need consists of a compression gauge and a spark plug socket, with the proper driver. Using a regular deep socket will probably cause a broken plug either on removal or replacement.

Bring the car to operating temperature by running it for 15 to 20 minutes. Remove any road dirt from around the plugs and then loosen them about one turn to break free any carbon. Start the engine and accelerate to 1000 rpm to blow out the carbon. Stop the engine and remove the plugs, placing them—in order of their removal—in an empty milk carton, electrode end up. Block the throttle in the wide open position. Remove the coil wire from the center of the distributor cap and ground it to the block.

If the compression gauge is the screw-in type, you can probably do this test yourself. If it is the hand-held type, you will need a helper or a remote control starter switch to crank the engine. Insert the tester in a spark plug hole. Crank the engine three or four turns and observe the reading. Write the reading down and repeat the test for all cylinders, recording the compression of each. Compare the readings. Each should be within 10 to 15 percent of the recommended compression pressure. Compression on six-cylinder engines should be 130 psi and 150 psi on the two-barrel, 265-cubic-inch V-8. It's better to have all the cylinders close to one another than to have a variety of highs and lows.

Readings higher than recommended are usually due to carbon build-up in the combustion chamber. Low readings might be due to either leakage past the rings or the valves. How to tell? Use the wet test. Squirt about a spoonful of engine oil on top of the pistons in the low-reading cylinders. Crank the engine a few times, and then take the compression test again.

If the reading is the same when compared with the first test, the valves are probably leaking. If the reading has risen about uniform with the other cylinders, it indicates compression loss past the pistons and rings. Should a low compression reading on two adjacent cylinders be obtained, it indicates the possibility of a leak from one cylinder to the other due to a "blown" head gasket. To check this out, torque the cylinder head down and take a compression test again. If any improvement is indicated, then the gasket might be the cause of the loss of compression.

ENGINE VACUUM

The engine vacuum test can be used in combination with a compression test. It will indicate the condition of the valves, rings, and manifolds. Its reading will show you what repairs you should consider on the valve train. Connect the gauge to the intake manifold and it will determine the difference between atmospheric pressure and the pressure in the manifold. Remember that there is a partial vacuum in the intake manifold as each piston moves down on its intake stroke. The better the seal the better the engine.

A normal engine, when idling, will give a steady reading between 18 and 22 inches of mercury. Most gauges are calibrated this way. When the throttle is opened and closed quickly, the needle should drop to around 2, then climb to 25, before returning to normal. Readings will vary with altitude and atmospheric conditions. Every 1000 feet of altitude lowers the reading by 1 to 2 inches of mercury. The V-8 engine will show higher readings than the six cylinder engine.

When the needle drops regularly by 3 or 4 inches of mercury, the cause is likely to be a burned or sticking valve. A similar reading can be due to a valve sticking in its guide. Rapid fluctuation of the needle is often due to loose valve guides, whereas wide fluctuation at higher engine speeds may be caused by weak valve springs. There may be several causes for each of the readings so recheck each one.

VALVE CLEARANCES

If the compression and vacuum readings are normal, the next step is to listen to the engine. Excessive tappet clearance will cause noisy operation. Normal engine wear will increase the amount of clearance on the early in-line six and V-8 engines that used solid-valve lifters. Engines with hydraulic lifters seldom require adjustments. A steady clicking noise in the valve train is one problem, but too little valve clearance is another that can lead to valve burning and loss of power under load.

Before you start, the engine must be warmed up so that all the valve operating mechanism has normalized. Let's do the 6-cylinder and V-8 engines with standard transmissions because they are the models that use the solid-valve lifters.

On the six-cylinder model, remove the rocker arm cover and gasket. You might be able to reuse this gasket but don't count on it. Have a new gasket available. Tighten the manifold center clamp bolts 15 to 20 foot pounds and the end clamp bolts 25 to 30 foot pounds. The cylinder-head bolts need 90 to 95 foot pounds in the proper sequence and the rocker arm shaft support bolts will accept 25 to 30 foot pounds.

You will need a screwdriver and a wrench to do the adjustment. It is recommended that you run the engine at a slow idle, but you can do a good job with the engine off by following this procedure. Chevrolet specifications indicate .006 of an inch for intake valves and .013 of an inch for exhaust valves. For longer valve life, consider 0.10 of an inch for intake and .020 of an inch for exhaust. You won't be sorry. It's a bit noisier but not so much that you would notice it.

The firing order of the six cylinder is 1, 5, 3, 6, 2, 4 and the valves can be adjusted at each cylinder by following this order. The piston will be at the top of the stroke on compression and both valves will be closed. Remove the distributor cap and note the location of the rotor and whether the points are starting to open. Adjust for the indicated cylinder using a feeler gauge. Crank the engine over to the next firing position and adjust these two valves. Continue setting the remaining valves by following the firing order. Replace the gasket and the rocker arm cover. Start the engine and check for oil leaks.

The V-8 engine is not much more difficult to adjust. The feeler gauge you will need is .006 of an inch for intake valves and .016 of an inch for exhaust valves. Normalize the engine and remove the valve covers. Torque the head bolts 60 to 70 foot pounds in proper sequence. You will need a head-bolt wrench adapter to reach some of the head bolts. The exhaust and intake manifold bolts torque to 25 to 35 foot pounds.

Crank the engine until the number 1 cylinder is on firing position, both valves closed, and the mark on the harmonic balancer lines up with the center or "O" mark on the timing tab.

Adjust the following valves: exhaust 1, 3, 4, 8; intake 1, 2, 5, 7. Crank the engine one full revolution until the mark on the harmonic balancer again lines up with the "O" mark on the timing tab. Now adjust these valves: exhaust 2, 5, 6, 7; intake 3, 4, 6, 8.

Install rocker arm covers, using new gaskets, and tighten the screws just enough to compress the gaskets. Any more and the gasket will split (causing a leak). Start the engine and check for oil leaks.

Hydraulic lifters are used in models having the Powerglide transmission (both 6 and V-8 engines). The lifter operates by automatically compensating for clearance. The biggest enemy of hydraulic lifters is dirt, and the greatest favor you can do your engine is to keep the oil and oil filter changed on a regular basis. An engine with many miles on it will probably have a noisy lifter or two, and especially when the engine is cold.

Lifters need an initial adjustment but you are not setting any valve clearance. They do not need regular adjustment, only when the valve mechanism has been disturbed (either by changing the rocker arms or exchanging the lifters).

On the six-cylinder engine, you can make the adjustment using the same procedure as you did for solid lifters. With the necessary piston on top dead center, turn the adjusting screw down until *all* lash is removed from lifter to valve. When you can feel no more side play of the push rod, turn the adjusting screw 1 1/2 turns more and securely tighten the lock nut. Adjust both lifters and continue to the next cylinder in the firing order. For the V-8 engine, tighten the rocker arm stud nut 3/4 turns more after *all* lash is removed from push rod to rocker arm. Use the same procedure as for solid lifters.

To replace noisy hydraulic lifters, you will first have to determine which one(s) are at fault. Remove the rocker arm cover on the offending side. With the engine running, place a finger on the face of the valve spring retainer. If the lifter is not functioning properly, a distinct shock will be felt when the valve returns to its seat. Sometimes a noisy lifter can be quieted by adding a concentrate to the oil in the oil pan. You could even try a half quart of transmission fluid. If the noise goes away, don't push your luck. Change the oil and the filter and the problem might remain solved. Generally, if the noise has gone on too long, you will have to replace the lifters that are causing the problem. Recheck and identify the location of the noisy ones. Remember that the engine compression is good. I'm not talking about an engine that is worn out and ready for a complete rebuilding.

Drain the radiator and remove the air cleaner. If you have some rubber-backed carpet around make up some fender covers. You don't know the damage a belt buckle can do until after it's done. Disconnect the throttle rod from tne carburetor. On Powerglide models, disconnect the lower transmission throttle lever rod from the bellcrank mounted on the coil bracket. Leave the carburetor on the manifold, but disconnect the fuel, vacuum and manual choke connections. Remove the water outlet hose and the heater hose.

With the distributor, take your time and you will get the engine running without any problems. Go ahead like a take-it-aparter and it's going to take you much longer to get the engine running. Use masking tape and identify the position of the spark plug wires. You could use a numbering system of your own, but the Chevrolet engine cylinders are numbered 1-3-5-7 on the left bank and 2-4-6-8 on the right bank (with you in the drivers seat). Pull the spark plug wires off the plugs and remove the distributor cap. Don't pull the wires out of the cap. Disconnect the distributor-to-coil wire from the coil terminal. Remove the vacuum line from vacuum control unit. Mark the position of the rotor on the distributor housing. Use a scribe rather than a pencil or chalk. Remove the distributor clamp bolt and the clamp. Note the relative position of the vacuum control unit to the intake manifold, and slowly pull the distributor up and out of the block. The rotor turns because of the helical drive gear on the distributor shaft.

Remove the bolts holding the manifold to the cylinder heads. Remove the manifold. Back off the rocker arm nuts at the lifters that are to be removed, and pivot the arms away from the push rods. Remove the push rods and the lifters; keep them in order. They have a wear fit so why change it? Use a clean milk carton for each set.

You can service the noisy lifters, but this is a time consuming job and it's very difficult to get parts if you need them. Examine the bottom of the lifter body. If it is worn or pitted, the entire lifter should be replaced. Examine the surface of the camshaft. If it is similar, you have a problem that needs correcting. If the lifter is noisy because of varnish or carbon that is causing it to stick, you could disassemble it and soak it in carburetor cleaner. Rinse with solvent, blow clean with dry air, and reassemble after checking all the internal parts for wear. If you are cleaning more than one lifter do not mix up the parts; they are not interchangeable. On assembly, use engine oil to fill the plunger. It's a lot easier to replace the lifters with new ones than try to service the old ones, but it's your time and money.

Check the push rods for straightness and that the oil passage is clear. Lubricate the lifter body and replace it in its respective bore and with the proper push rod. Pivot the rocker arm to engage the push rod and tighten the rocker arm nut to about where it was.

Install the manifold with new gaskets and torque in the proper sequence. To install the distributor, first turn the rotor about 1/8 turn in a counterclockwise direction, past the mark you scribed on the distributor housing. Push the distributor down into position in the block and the rotor should line up with the scribed mark.

Tighten the distributor clamp screw and connect the vacuum line and the coil wire. Install the cap and place the spark plug wires on the proper spark plugs. It's not difficult at all when the wires are marked. Now, you can crank the engine over to adjust the hydraulic lifter(s) as required. Use the procedure listed earlier.

Connect all other lines, linkages, and hoses. Fill the radiator. If the lifter(s) weren't filled with oil during servicing, start the engine and let it idle until the lifters are all pumped up. Adjust the lifters, replace the rocker arm covers, use new gaskets (if necessary) and you're ready to roll.

To service hydraulic lifters on the six-cylinder engine is a lot quicker and easier. Remove the rocker arm cover attaching screws and remove the cover and gasket. Install a new gasket if this one cannot be reused. Silicone is a great help. Disconnect the spark plug wires (knowing that the firing order is 1-5-3-6-2-4) and remove the center coil wire. Remove the push rod cover attaching screws and remove the cover and gasket. If the gasket can be reused, that is fine. If not have another one ready. Silicone doesn't solve everything; ask the new car mechanics.

Remove the bolts and nuts that retain the valve rocker arm assembly to the cylinder head and remove the rocker arm assembly. Depending on which lifter(s) are noisy, you might be able to loosen the adjusting screw in the rocker arm enough so that the push rod will come out. Then remove the lifter and replace it or service it. Replace the pushrod, adjust lifters as previously discussed, and you have saved some time. Install the pushrod cover and gasket, the rocker arm cover and gasket and you're on the road again. The spark plugs need wires.

VALVES, RINGS, AND BEARINGS

The easy things are done. You know the condition of the engine—through compression and vacuum gauge tests—the valves are set, and now you want to do a bit more. The "wet" and "dry" test indicates a bad valve(s). The condition of the valves, more than anything else, determines the power, performance, and economy of a valve-in-head engine. In other words, you are going to have to do something to bring this engine back to its original performance.

What you do will depend on your experience, knowledge, and the amount of money you can spend. It also depends on whether you have another car to drive while you repair this one and if the necessary tools are at hand. You can work under the shade tree but you cannot work without tools. They are but an extension of yourself. Specialized tools are best left to the pros. If you can get the head(s) off, the machine shop will do the rest. They are the specialists and they have the tools.

Before you start anything, raise the hood and get a good mental picture of what you want to do. Look at everything that is connected to the head(s) and what must be disconnected to take the head(s) off. There isn't that much difference between the V-8 engine and the six (more parts of course). The general lay out is similar. Valve-in-head engines are called overhead valve engines and you can see why.

Drain the coolant and disconnect the battery ground clamp on either engine model. Cover the fenders, and while the engine is cooling remove any of the parts that are attached to the head. Removing a cylinder head while hot could cause it to warp upon cooling.

On the six cylinder, remove the air cleaner, all rods, lines, and vacuum tubes at the carburetor and manifold. Unbolt the manifold from the cylinder head, but not away from the exhaust pipe flange. Leave the carburetor on the intake manifold. Pull the connected manifolds away from the head. Now, go to the other side and remove the push rod cover and the rocker-arm cover. Remove the gas and vacuum line retaining clip from the water outlet. Disconnect the water outlet and remove the thermostat. Remove the temperature indicator element. Don't kink or bend the line. Disconnect the oil line leads to the rocker arms and remove the four bolts and two nuts that retain the rocker arm assembly to the head.

Remove the rocker arm assembly. Keep it in one piece so you can slip it back on the studs on the head. Remove the cylinder-head bolts and the cylinder head. Place the rocker arm assembly on the head and secure it with the two nuts.

Remove the intake manifold sleeves (if they are in the head) and place them in the intake manifold. Make an inventory of what is on the head. This is important because you are not the only machine shop customer and parts can get lost. You can transfer this list to the work order and there will be fewer problems on delivery day. Deal with a machine shop and employees that show a little interest toward the '50s iron.

While your waiting, there are a few things you can do. Check the valve lifters (solid or hydraulic); they must be a free fit in the block. Carbon or varnish should be cleaned off. If there was a noisy lifter, this is the time to replace or service it. The ends that contact the camshaft should be smooth. If this surface is worn or rough, the lifter should be replaced. The pushrods must be straight; roll them on a piece of flat glass to check for a bent condition.

Check the manifold heat control valve shaft; it must be free in the manifold. Free it up using heat riser solvent or kerosene. Careful tapping back and forth with a hammer will also break the carbon loose. Don't oil the shaft. The spring tension should require only half a turn from the unhooked position to slip it over the anchor pin. This is a thermostatic spring. Replace it if distorted in any way.

Clean all the pressed steel parts and spray them with the proper shade of blue to prevent rusting. Clean the screw holes in the block using a suitable tap. Blow the holes clean. Wash all the bolts and use a wire brush on the threads. Replace any that are stretched or checked. Obtain a valve grinding gasket set and you are ready for assembly. Check for cylinder ridge by using your fingernail. If you can hook it against the ridge, some wear is indicated.

The "wet" "dry" test also is a good indicator of ring and cylinder condition. You know how much oil the engine is using, and if it's not leaking out the engine is burning it. Rest assured that doing the valves will not cure an oil-burning problem.

Place the new cylinder-head gasket on the block by following the suppliers instructions. Check the water passages and bolt holes for proper alignment. If you have two old head bolts, saw off the heads and cut a screwdriver slot in place so that you have guide pins to keep the gasket in alignment with the head. Carefully place the head in position and lower it over the guide pins. You can also make a pair of "T" handles by drilling a 3/8-inch hole in two short pieces of 1-inch pipe. Place these handles over the studs in the head and run the nuts down. This makes handling the head just a bit easier.

Install all the head bolts finger tight and then torque them in proper sequence to the final torque of 90 to 95 foot pounds. Install the manifold using new gaskets and inserting the intake manifold sleeves. Tighten the center clamp bolts 15 to 20 foot pounds and the two end clamp bolts 25 to 30 foot pounds.

If the lifters were serviced or replaced, install them in the block. Install pushrods and the rocker-shaft assembly. Connect the oil line and insert the temperature-indicator element. Adjust the valve clearances. Install the pushrod cover, using a new gasket, and tighten attaching screws evenly to 6 to 7 1/2 foot pounds of torque. Connect the water outlet. See that the thermostat and gasket seat properly to prevent coolant leaks. Connect the gas and vacuum lines to the clip. Add the coolant and check the oil level. It's not a bad idea to change the oil and filter.

Adjust spark plugs to .035 of an inch, put them in place, and attach the spark plug wires in the correct firing order 1-5-3-6-2-4. Start the engine and check to see that the rocker arm shafts are getting oil and that there aren't any noisy lifters or valve clearances. Use a new rocker arm cover gasket and insulators and replace the cover. To goo or not to goo; it's up to you. Start the engine check for oil and coolant leaks. If everything is fine, you are ready to roll.

With the V-8 engine, you have many parts. Make diagrams as to where the parts fit. Memory is pretty short on things we have very little experience with. If only one valve is indicated as burned, you might be able to service only that head, assuming the compression is up to specifications on the other. Use the test equipment you have; compression gauge, vacuum gauge, and the "wet" "dry" test, and you will be able to analyze the engine problems.

Oil consumption can be caused by leakage or burning internally. Leakage can be corrected by replacing gaskets or seals. Burning the oil means that it is getting into the combustion chamber, either past the rings or past the valve stem seals, and the valve guides. Seals and guides are serviced with the valves. The machine shop does this work. If you find a cloud of blue smoke immediately after acceleration and after idling at a stop light, you can assume that oil is getting past the valve stems and into the combustion chamber. If the valves are in good condition you can replace the seals only without removing the heads. Use umbrella or Teflon seals and not the regular rubber O-ring type.

Do not attempt this service until you obtain a valve spring compressor that will work with the cylinder head still mounted on the block (GM # J-5892). Remove the valve covers and rocker arm shafts where necessary. Remove the spark plugs and use a screwdriver to hold the valve in place or insert a length of clothesline. Turn the engine by hand to bring the piston to the top. Compress the valve spring with the compressor, remove the keepers, and release the tool slowly. It's under spring tension so be careful. Remove the valve spring and retainer. Remove the old seal. Follow the manufacturer's instructions for installing the new seals. Install the spring and retainer and compress the spring. Reinstall the keepers and make sure they fit in the grooves. Release the tool and so on to the next valve. When you are finished adjust valves as noted before. Replace cover(s).

This might be all the work that you want to do on the engine. If you are still adding oil between oil changes and if this oil is burned, you have a problem. Do a compression test again ("wet" and "dry") and analyze the results. Is the oil getting past the rings? Okay. This is the expensive part. If the engine needs a rebore and new pistons, you should consider having the engine completely overhauled. Do one more test before you make a final engine decision. Remove the oil pressure switch from the block and replace it with a pressure gauge. Depending on how much a purist you are, you might already have changed the so-called idiot light with a direct reading gauge. The gauge will give you a good idea of the internal condition of the engine bearings. At 30 miles per hour and at operating temperature, the gauge should read 30 psi. The light operates at 5 psi and over so you can see that it is not a good indicator of engine-bearing condition. Remember that any engine with over 100,00 miles is starting to suffer from old age and a rebuilding job will have to be done to restore it to its original performance.

How much you can do depends on how much you want to do. First and foremost—if you are not comfortable—leave the engine alone. Its easier to go the exchange route and it's a job you can do in a weekend. If the tests indicate that replacing rings will solve the oil-burning problem, then do it. Also examine the bearings that support the crankshaft and those that are in the connecting rods.

Collect or rent the following tools before you start. If this is going to be a week-ender, have the gaskets and parts at hand. When the engine is disassembled, you might find it has been rebored or the crankshaft has been reground and standard parts will not fit.

It's a good idea to find an auto parts store that is open on Saturdays.

You will need a torque wrench, cylinder ring ridge remover, glaze buster ring compressor, flat feeler gauge, and a few strips of Plasti-gauge that you can buy at the auto parts store.

Remove the head(s) as under valve service and clean one of the piston heads with a wire brush. Look for any numbers stamped on the piston head such as 010, 020, 030, 040. These will indicate oversize pistons that in turn mean oversize rings. Raise the front of the vehicle on stands and drain the oil. Disconnect the steering idler arm to allow the oil pan to clear. Unbolt the oil pan retaining bolts and lower the oil pan. Check for stamping marks on the crankshaft counter weights (010, 020 indicate undersize bearings).

Use the Plasti-gauge as per instructions and check the clearance of the connecting rod bearings. Torque 35 to 45 foot pounds. With clearances greater than .004 of an inch, you might consider changing the bearings if the crankshaft is not scored or out-of-round. Use the cylinder ridge remover and ream out the ridge. Push the pistons out of the block. Note the position of the numbers stamped on the rods and caps. Try not to nick the crankshaft with the rod bolts. You can use short lengths of rubber hose on these bolts to prevent marring the crankshaft. Replace the caps on the respective rods. You can buff the piston heads but do not buff the piston skirts. Remove the piston rings and clean the grooves by making a chisel scraper out of a piece of ring. Don't remove piston material—just the carbon.

Clean the oil-return holes. Check the cylinders for deep scores and use the glaze buster to get a cross-hatch pattern on the cylinder walls. A portable drill will operate the glaze buster satisfactorily. Wash the cylinders with hot soap and water to remove any grit. The grit shortens ring life so the cleaner the cylinder the better.

Remove the oil pump. On the six cylinder a locking bolt and nut by the camshaft bearing hold the pump in place. On the V-8, a bolt holds the pump to the rear main bearing cap. On both models, a tang on the distributor shaft drives the pump. Clean the oil pump screen in clean solvent. Remove the cover and slide off the idler gear. Clean the inside of the pump body. Examine the cover and the pump body for scoring or deep scratches.

Replace the idler gear and check the clearance between it and the drive gear with the feeler gauge (.003"-.004" maximum). The clearance around the outer rim of the pump body and either of the gears should not exceed .003 of an inch. The inside of the pump cover can be cleaned up by surfacing it on a piece of flat glass and emery paper. See that the relief valve is free in its bore. If you have any doubt about the pump or the pressure it develops, replace it with a rebuilt one or a new one. This is the most important part of the lubrication system.

Check main bearing clearance using Plasti-gauge. Do one bearing at a time and torque back up into place (100 to 110 foot pounds for the six cylinder and 60 to 70 foot pounds V-8. Maximum clearance is .004 of an inch. If the clearance is above this limit on the six cylinder, remove shims as necessary. If clearance cannot be adjusted, the crankshaft will have to be reground and fitted with new undersize bearings. Install new rear bearing oil seal.

On the V-8 engine, it might be possible to use a .002 of an inch undersize bearing to produce the necessary clearance. If not, the crankshaft will have to be reground. This engine does not use shims. Replace the rear main bearing seal. Any bearings that show fatigue, abrasion, or scoring should be replaced. Follow the instructions that come with the rings, they will tell you everything. Leave each ring in the package until you are ready to install it. Install the bottom oil ring first. The rings must be free to move in the groove. Hold the rod in the vise (not the piston). Install the compression rings with the gaps 180 degrees apart and 90 degrees from the piston pin ends.

Dunk the piston and rings in clean oil, put on the ring compressor, and tighten it down. If you are replacing the connecting-rod bearings, don't fingerprint the bearing surface. Slip on the thread protectors. Coat the cylinder walls with engine oil and rotate the engine for the particular piston so that crank journal is at bottom dead center (BDC) position.

This is a good place to have a helper because one person can be under the car guiding the bolts past the crankshaft journal. Set the piston in place, ring compressor tight against the block, and drive the piston down using the hammer handle until the top of the piston is flush with the block and the ring compressor is free. Slide under the car and pull the rod down until the bearing surface is tight against the journal. Oil the cap bearing and place the cap on the rod (numbers same side). Tighten the nuts snug, but leave the torquing until all cylinders are filled. On the six cylinder, the running mates are 1-6, 2-5, 3-4. Do the one that has the crank journal at the bottom. On the V-8, the rods are paired on the crankshaft journal so do the corresponding one. Oil the bearings well to prevent any dry starts. Torque the rod bolt nuts 35 to 45 foot pounds. Rotate the crankshaft at least one full turn after final torquing to be sure that the bearings are not too tight.

Install the oil pump. Assemble the collar end of the extension shaft assembly over pump drive shaft, aligning tang of extension shaft with slot in end of pump drive shaft. Install retaining clip with flat end of clip in groove of pump drive shaft. Install oil pump and extension shaft assembly to rear main bearing cap, and torque cap bolt 45 to 50 foot pounds. Don't leave out the extension shaft because it drives the oil pump via the distributor.

On the six cylinder, place the oil pump in position in the block fitting. If the distributor is in place, align the tang with the slot in the end of the pump drive shaft. Tighten the oil pump retaining screw. Be sure that the tapered end of the screw draws down into the hole in the pump body. Tighten the lock nut securely. Install the oil pump to block the oil line and tighten the connector nuts.

Install new oil pan gaskets and end seals. Use sealer to hold gaskets in place. Carefully raise the oil pan into position and start all the bolts in place. Tighten corner bolts 12 to 15 foot pounds and flange bolts 6 to 9 foot pounds. Tighten the oil pan drain plug. Change the oil filter if it is located at the bottom. Reconnect the steering linkage and lower the vehicle. Put the oil in now. If you have a suitable driver for an electric drill, prime the oil pump. If you had the head(s) serviced, replace them now. Attach the manifolds. Use new gaskets and torque as required. Install the rocker arm assemblies and adjust the valves. Install the thermostat housing and thermostat. Time the distributor and engage the pump drive. Install a wiring harness and attach it to proper spark plugs. Connect the carburetor. Connect the electrical wiring. Add coolant and start the engine. Immediately shut off the engine if noise develops or the oil light stays on.

Remove the oil pressure switch and replace it with a gauge. Start the engine and observe oil pressure. If the pressure is acceptable replace the switch. Noise may be due to something that you did or did not do. Get the opinion of a professional mechanic on this one.

If you decide to go the exchange route, you should be able to handle this job on a weekend with a little help from your friends. You will need a chain hoist or a block and tackle to lift the engine out of the vehicle. Remember the supporting structure must be strong enough to handle the weight. Don't pull the roof of your garage down on yourself. Another way is to rent a tow truck with a boom and let the machine lift the old engine out and the new one in.

Prepare for lifting by removing the hood after scribing the hinge position. Set it aside where it won't get damaged. Drain the oil and coolant and remove the radiator and mount. This way you have plenty of room to remove the engine and the transmission as a unit. Disconnect front and rear engine mounts. Split the rear universal joint and slide the drive shaft back until it is free of the transmission. Tie a plastic bag over this opening to prevent loss of fluid. Disconnect the linkages and lines (whether standard or Powerglide). Disconnect the speedometer cable. Disconnect the battery ground strap and connections from the starter. Disconnect the exhaust pipe from the manifold(s). Disconnect the rear transmission mount. Remove the starter.

Remove the fan blades, hub, and water pump. Take off the air cleaner and remove the carburetor. Pull out the distributor and remove the high-tension wires. Disconnect the primary wires and remove the coil. Disconnect the generator leads and remove the generator. Mark all wires with masking tape to identify their origins. Disconnect the temperature sending unit and oil pressure unit.

Remove the choke heat tube and rocker arm covers. Attach a good lift chain to the cylinder head bolts on the V-8 or 6 engine. Remove the engine and the transmission out of the vehicle. Use caution. Don't get under this unit. As soon as it is clear of the engine compartment let it down gently on a suitable support stand. This is a heavy unit. If you are using a set of saw horses and short planks, nail them together. Set the engine on the floor if you must. It's a bit inconvenient but much safer.

To remove the standard transmission, you have to remove the clutch pan and the bolts that hold the transmission to the clutch housing. Support the transmission as you pull it away. This will prevent damage to the clutch disc. Remove throw-out bearing from clutch fork. Loosen the clutch bolts a little at a time to prevent distortion of the clutch cover. Remove the plate and disc. Remove the flywheel and clutch housing.

On the Powerglide model, remove the flywheel inspection hole cover and turn the engine over until one of the three flywheel-to-converter-attaching bolts appears in the opening. On the V-8, this opening is by the starter mounting. On the six it is on the left side of the engine. Remove all converter-housing-to-flywheel-housing attaching bolts. Slowly move the transmission back. It's plenty heavy so support it. Keep the converter on the transmission and set it out of the way.

Remove the flywheel and the flywheel housing, if there are any shims here make a note of their location.

Comparing the two engine assemblies, it's easy to see what you will have to remove. Clean all the parts, paint them if you like, and start assembly. It's basically a reversal of disassembly. When bolting the clutch housing or flywheel housing back in place, make certain there are no burrs on either mounting surface. Bolt torque is 25 to 35 foot pounds. Make the same check on the mating flanges of the flywheel and crankshaft. Bolt is torque 55 to 65 foot pounds. Check Chapter 8 for transmission mounting.

Lift the assembly back into place, align mounts and tighten down. Remove the lift chain and retorque the cylinder-head bolts. Leave the rocker arm cover(s) off the case the valves need readjusting. Connect all linkages and wires. Replace coolant and lubricant. Start the engine and normalize the temperature. Make any adjustments and replace the rocker-arm cover(s).

Use automatic transmission fluid (Type "A") if the fluid level is low. Do not overfill. Replace the hood assembly, and you are off for a test drive. Feels good doesn't it. You have had the pleasure of bringing a golden oldie back to its original performance.

Chapter 4

The Electrical System

There is no magic in an automotive electrical system. Electrons want to flow from negative to positive. The battery has one positive post and one negative post. The negative post has the surplus of electrons so the current flow is from negative to positive when there is a complete circuit. The voltage forces the electrons through the circuit when you close any of the switches and the system works. This particular system uses a 12-volt battery with a negative ground connection.

Very little can go wrong with the electrical system. If the battery does not crank the engine it can be recharged or replaced. The wiring that makes up the circuits can either be open, shorted, or have high resistance. An open is a break in the circuit. It can be caused by a broken wire, an unplugged connection, or a burned fuse. A short is when the circuit takes a "shortcut" to ground. High resistance is caused by corrosion or a bad connection. You can fix any of these problems by following a check-out procedure that is thorough and orderly.

It's nice to have some test equipment, but common sense is one of your most important faculties. Use it. The battery is the heart of the electrical system and the first place you should check. A battery hydrometer is an excellent addition to your tool inventory. An ammeter and a low-reading volt meter or a voltohmmeter (VOM) are handy. A 12-volt test light makes a great continuity tester. Follow the manufacturer's instructions and you can save yourself some money, don't invent shortcuts; it's going to cost you money.

BATTERY AND STARTER SERVICE

If the battery doesn't crank the engine don't blame the starter. Instead look at the battery. Examine it for corrosion, cracking, and leakage. Inspect the hold-down, posts, cables, and electrolyte level. The electrolyte is made up of sulfuric acid and water and will cause serious burns to the skin and eyes. Flush with plenty of cold water. Battery acid on the car finish or on clothing should also be flushed with cold water and neutralized with a mixture of baking soda and water. Never strike a spark, light a match, or bring an open flame near a battery. The charging process creates a mixture of hydrogen and oxygen gases that can ignite and burn with explosive force. This could blow the case apart, spraying acid over a large area. Don't smoke.

Corrosion causes high resistance at the posts and cable ends and it should be cleaned away. Use a wire brush to remove the heavy stuff. Remove the battery cables. Use a puller if they are tight (to prevent damaging the battery). Cover the cap vent holes with masking tape. Brush a solution of baking soda and water over the terminals, posts, and battery top. Let the solution work until the foaming stops; add more if needed. Rinse with water and wash this mess away from the rest of the car and concrete or driveway. Wipe dry and clean the posts and the inside of the terminals with sandpaper. Coat with nonmetallic grease and tighten securely. Remove the masking tape.

The hold-down should be tight and the top of the battery should be dry. A dirty top attracts electrolyte and will cause slow battery discharge. Do a hydrometer test and measure the specific gravity of the electrolyte before adding any water to the cells (if they are low). The level should be above the top of the plates by 3/8 inch. Correct for temperature and write the readings down.

If they are between 1.215 and 1.270 and the variation between cells is less than 25 gravity points, the battery is in good condition. If below 1.215 but the variation is acceptable, then the battery should be recharged. A trickle charger will do the job. If specific gravity readings show a variation of more than 25 gravity points, the battery is worn out and should be replaced. This test does not indicate whether the battery will start the vehicle. It is a test for determining the state of charge.

A capacity test is required and a voltmeter is necessary to do this test. Connect the voltmeter and ground the coil high-tension lead. Crank the starter for 15 seconds and observe the voltmeter reading; it should not be less than 9.0 volts. If the reading is more, the battery is fine. If the specific gravity reading is above 1.215, no service is required. Check the fan belt to see that it is not loose.

A battery that continually needs water is being over charged, and the voltage regulator should be replaced or adjusted. Occasional topping up of battery electrolyte should be done with distilled water. Tap water with high mineral content should not be used. Sealed, maintenance-free batteries do not require that water be added.

With the battery in good condition, the starter should engage, crank the engine fast enough to start, and disengage when the engine starts to run. If the starter motor runs but does not crank the engine the problem is in the drive unit and the starter will have to be removed for servicing. A slow cranking speed indicates excessive resistance at the starter. If the solenoid chatters but does not engage the starter, then it should be replaced. You cannot inspect this starter without removing it because it does not have an inspection band.

On the V-8, raise the car on stands and don't drop the starter on yourself. On the six, you can remove the starter from the top. Remember that it is heavy. Don't remove the last bolt without being ready to remove the starter. Disconnect the ground cable at the battery and then the hot cable at the starter. Identify any of the solenoid wires. Wash the outside of the starter and disassemble. Wash the parts in solvent except the armature, field coils, and drive unit. Inspect all the parts and have an electrical shop test the armature and the field coils.

If either of these are unsuitable, buy an exchange starter and put it in place. The cost of piecing a starter together is not worth it when one or both of these parts are nonserviceable. If the commutator is rough or out of round, it should be turned down and undercut.

Check the bushings in the drive housing and the commutator end frame. The shaft should fit snugly so that the commutator does not rub on the pole shoes. Look for evidence of this. The drive unit should turn freely in the overrunning direction, but not in the cranking direction. Check the teeth for cracking or chipping. Replace as necessary. Check the brush holders to see that they are not deformed or bent, but will hold the brushes against the commutator. Replace brushes that are shorter than 3/8 inch.

If the windings in the solenoid are drawing the proper

amperage, you can replace or clean up the motor connector strap terminal and turn the contact ring over to make a better connection. Place the solenoid in position on the starting motor and install the four mounting bolts. Install the solenoid plunger and connect the linkage to the shift lever. Adjust the pinion clearance by pushing the plunger in as far as it will go and hold it in place. Push the pinion on the drive unit as far back as possible and with a flat feeler gauge check the clearance between the end of the pinion and the pinion stop (with pinion in cranking position). Clearance is .010 of an inch to .140 of an inch. Adjust by loosening the screw in the plunger linkage and shortening or lengthening the linkage as required. Retighten the screw securely when the adjustment is correct.

Try the starter before you mount it in place by attaching some booster cables and giving it a free-run test. It should reach a speed of 6900 rpm. Replace starter and attaching bolts or stud nut. On the six cylinder, an engine-to-body ground strap is attached to the bottom-mounting bolt. Reattach the wires and cables. There is no place for lubrication on this starter after it is assembled. Do it before. Use 20-weight oil but avoid excessive lubrication.

IGNITION SERVICE

If the engine cranks but does not run, the ignition system might be the problem. The ignition system consists of a coil and resistor, distributor and cap, rotor, ignition points, condenser, spark plugs and ignition switch, together with the battery and the necessary low-and high-voltage wiring. There are lots of connections and lots of places for resistance to cause a problem. Any loose connection will cause an open. If the engine will start but immediately stop when the ignition switch is released from the start position, then replace the resistor.

You can check the low-voltage circuit with the test light and the high voltage by pulling a spark plug wire off and holding it away from the block about 1/4 of an inch. Crank the engine over; there should be a spark. If there is a spark, the ignition system is not at fault. You better make sure the tank is full of gasoline and the fuel system is cooperating. Back to the no spark situation. Replace the spark plug wire and pull the center wire out of the distributor cap. Hold it about 1/4 of an inch away from the block and crank the engine. If there is a spark, the problem is in the high voltage (either the rotor, cap, or high-tension wiring). If there is no spark, the problem can be in the distributor, coil, or related wiring.

Use the test light and see if the current is coming to the coil primary terminal and going out the distributor side when the points are closed. If it isn't have the coil checked. The coil is a step-up transformer and changes the battery voltage of 12 volts to 20,000 volts to jump the gap at the spark plug.

Remove the distributor cap and examine the points. They are satisfactory as long as the voltage across them does not exceed .125 volts. Use a voltmeter across the points, with the points open and the ignition key on. Do a visual check, the points should not be burned or pitted; if they are replace them. Replace the condenser also. You can do so with the distributor in place. Use a screw starter to make the job easier. Replace all insulators in the same position you found them. Point opening should be set at .019 of an inch, using a flat feeler gauge. Check that the distributor lead is connected to the negative side of the coil and that all low-voltage connections are clean and tight.

Take the distributor cap with the wires attached. Pull one wire at a time and examine the tower for corrosion, burning, or flashover. Flashover is caused by the high voltage taking a short cut to ground and burning the cap. It can be prevented by having tight, clean connections and keeping the top free of dirt and moisture. Corrosion can be cleaned with fine sandpaper. Inspect the inside of the cap for flashover and cracks. If the terminal posts are grooved or burned, replace the cap. When replacing the cap, also replace the rotor. High tension wires that are burned or cracked should be replaced. See that the boots fit tight on the cap.

Spark plugs should be replaced at 10,000-mile intervals. If they are running too hot or too cold, change the heat range. Faulty plugs waste gas, cause hard starting and general poor engine performance. The gap should be .035 of an inch. There is no money to be saved by re-gapping and cleaning the old spark plugs. Use a proper spark-plug socket when removing and replacing plugs to prevent breaking the upper part of the insulator. Replace the spark plug wires on their respective spark plugs. Arrange them in the holders to prevent crossfiring. Start the engine and, if you can get hold of a timing light, check the ignition timing. Sometimes it will change.

CHARGING CIRCUIT SERVICE

The telltale light operates as an indicator of generator output and not whether the battery is accepting charge. If the light stays on after the engine is started and run above idle speed, the generator should be checked. Make sure the fan belt tension has about 3/4-inch deflection and that all the wires on the generator and regulator are tight. If you have a tendency to over oil the generator, it's possible that the commutator and brushes are not making contact. You could spray some grease-cutting product through the openings in the generator end frame, with the engine running, and see if the light goes out.

If you have an ammeter or a voltmeter, make a few more tests. See that the regulator is firmly mounted and a good ground connection is evident. Remove the wire from the "BAT" terminal of the regulator and attach the ammeter leads to the wire and the regulator. With the engine running at above idle, use a screwdriver to ground the "F" terminal on the generator for an instant. This lets the generator produce full output and the ammeter should indicate this fact. If the reading is reversed, switch the ammeter leads. If the reading is not affected by grounding the "F" terminal, disconnect this lead and notice if the reading falls. If it does not, remove the generator and check for a grounded field.

Oxidized points in the regulator may also be the cause of low generator output or a discharged battery. Turn on the headlights and run the engine fast enough so that a reading of 5 amperes is indicated on the ammeter. Ground the "F" terminal of the regulator with a screwdriver. If the ammeter reading increases by more than two amperes, the regulator should be removed for cleaning. If the generator passes the above tests, unhook the ammeter and connect the wires. Connect the voltmeter at the "BAT" of the regulator terminal and to ground. Start the engine and note the voltage; it should not be over 14.8 volts. If it is, you have a thirsty battery and a regulator that needs replacing. High voltage will also damage lighting and ignition circuits. Replace rather than repair the regulator. If the battery and generator are in good working condition, then you will have a balanced electrical system.

The generator might be in good working condition electrically, but mechanically it may be noisy and the bearing should be replaced. Disconnect the wires and mark them with masking tape. Remove the generator brace and the fan belt. Remove the support bolts. The generator is fairly heavy so don't let it drop. Before you take it apart, give it the nose test. If you can smell burnt varnish or insulation, and you know the generator had no output, replace it with a rebuilt unit. The price of parts is always greater than the sum of the whole.

To replace the bearing, remove the nut and lock washer at the drive end. Slide the pulley and fan off the armature shaft and remove the key. Remove the two through bolts and lock washers and remove the commutator and frame. Examine the brushes. If they are worn to half their original length, they should be replaced. Remove the drive end frame and armature assembly. Slide the drive end frame and spacer collar off the armature shaft. Remove the spacer collar from the end frame and the spacer washer from armature shaft. Remove the ball bearing from the drive end frame.

Wash the bearing in solvent and check for roughness. Check the fit of the armature shaft in the bushing in the commutator end frame. If the bushing is excessively worn, the end frame should be replaced. Inspect the armature commutator. If it is rough or worn it should be trued and undercut. If it only needs cleaning, use No. 00 sandpaper and drive it with a low speed on a drill press. Blow it clean. If there is a no-charge condition or a grounded field was diagnosed, have the electrical shop test the components. After all parts have been inspected and worn or damaged parts have been replaced, the generator can be reassembled.

If the bearing was satisfactory, pack with a nonfiber, high-melting point grease till about half full. Install in the drive end frame after the felt washer has been lightly oiled. Assemble these parts in the correct order. If you are replacing brushes, insert them so that the taper end will correspond to the curvature of the commutator. You can seat new or old brushes by taping a strip of sandpaper to the commutator, grit-side out, installing the armature in the housing, and rotating the armature to seat the brushes. Blow the dust away, and remove the sandpaper.

Assemble the commutator end frame, position on dowels and install the through bolts. Assemble the key, fan, and pulley to shaft, and then install lock washer and nut. Spin the pulley by hand. It must be free to turn.

Check the generator before installing it by making it run as a motor. Connect the "A" terminal on the generator to the positive post of a 12-volt battery. Connect the negative post to the generator housing. Ground the "F" terminal and the generator should "motor." If you insert an ammeter in the positive side, it should read 5.6 amperes maximum. Remount generator in position and adjust the fan belt. Attach the wiring and on radio-equipped cars connect the condenser to the "A" terminal. Before you start the engine, polarize the generator to ensure the current will be traveling in the proper direction. Connect a jumper wire between the BAT and GEN terminals on the regulator for no longer than a second.

Give the generator a final oiling with a few drops of 20-weight engine oil.

LIGHTING AND WIRING

Two separate wiring harnesses—body wiring and chassis wiring harness—are provided in all passenger car models. Multiple connectors, located behind the left end of the instrument panel at the cowl, are provided to join the body and chassis harnesses. An optional accessory junction block is located on the upper inside of the dash panel to the left of the steering column. It provides power take-offs and fuse clips for all accessories. Do not attempt any electrical work without a wiring diagram. You can only trace the wires when you know where they are to go and what they are to do. This is made easy by color coding the wires through the use of plastic insulation.

The lighting system includes the lighting switch, stoplight switch, dimmer switch, sealed-beam head lights, park lights, tail and stoplights, and instrument and indicator lights. The lighting switch uses two circuit breakers. One protects the headlights and parking lights while the other protects all the other lights. Very little can go wrong with the wiring. Make an occasional check to see that the connections are tight, sockets are grounded, and the wiring is not chafed.

Get in the habit of doing a walk around with the light switch pulled full out so you can check the lights. Press the dimmer switch to change the lights from low beam to high beam. On high beam, a small red indicator will be visible through an opening in the speedometer dial. See that the stop-light switch works when the brakes are applied. The directional signal switch uses the parking light bulbs at the front and the stop light bulbs at the rear to light its way. Bulbs and sealed beam units do burn out, and that's the first check you should make.

To replace a sealed beam unit, remove the screws that hold the headlamp door in place. Sometimes this is called the head light ring. There should be a rubber gasket here. On the early models, the retaining ring is held in place by three screws, look carefully and don't disturb the horizontal and vertical adjustment screws. On later models, a spring holds the retaining ring. Use a pair of pliers to unhook the spring. Do not disturb the adjusting screws. Pull the sealed beam unit forward and disconnect the connector plug from the unit. On earlier models, connect plug to a new unit, place it in position, and install a retaining ring. Tighten screws and replace the headlamp door and screws. On later models, there is a retaining tab that holds the retaining ring to the mounting ring. Place the new sealed beam into the mounting ring, and then engage the retaining ring and connect the plug. Slide the mounting ring screw tabs into the vertical and horizontal adjusting screws. Pull the spring into the spring slot. Install the headlamp door with a gasket and secure it with the screws.

To replace the bulb in the park lamp or in the taillight and stoplight, remove the lamp door and the lens. The bulb should come out of the socket. It might be corroded in and it might not be the bulb that is at fault but the socket. When the socket is corroded, the housing might also be in the same shape. Try and get NOS and replace the housing. You might have to break the bulb to get it out, and then use a pair of pliers to remove the base.

Clean the corrosion with steel wool and use the test light to check for power. These same bulbs are the cause of turn signal problems. Check the socket, the wiring, and the ground connection. If park and stoplight bulbs are both on when signaling, check the flasher unit. When the taillight bulb and socket check out, but the stoplight does not come on when the pedal is depressed, the stoplight switch might be at fault. Use the test lamp. There should be power in both wires when the switch is depressed. The switch is under the instrument panel adjacent to the brake pedal.

Early models are threaded into the brace while later ones are retained by a second lock nut. Locate the new switch-on brace so that electrical contact is made when the brake pedal is depressed 5/8 of an inch from fully released position. Install the lock nut and the two wires, and check for operation.

The instrument cluster contains all the gauges and illuminating bulbs. Do not work under the dash until you remove the battery ground strap. The quickest way to have a fire is to short something out. Depending on how talented your fingers are you might be able to replace any of the bulbs by snapping the socket out. To replace the gauges, you will have to remove the cluster. Remove the upper and lower steering gear jacket covers. Remove two screws retaining the bottom of the cluster.

On Powerglide models, disconnect indicator rod from lever. Remove two screws under the lip of the instrument panel cluster hood. Pull the cluster carefully, toward the rear of car, out of panel opening. Replace gauges as required by removing attaching screws. To assemble, reverse the above procedure. Good luck.

Chapter 5

The Fuel System

The fuel system includes the fuel tank, fuel pick-up pipe and screen, fuel gauge, fuel lines, fuel pump, carburetor, air cleaner, intake manifold, and exhaust system. The fuel tank is located at the rear of the vehicle and holds 16 gallons of gasoline. In the case of a fuel-system problem, the first thing you should check is that there is gasoline in the fuel tank. Repairing the carburetor or replacing the fuel pump will not get the engine started if the fuel tank is empty. The fuel tank must be vented to the atmosphere to keep a constant pressure on the fuel. This pressure and the vacuum produced by the fuel pump deliver the fuel to the carburetor. The electric fuel gauge lets the driver know how much gasoline is in the tank.

The fuel lines transfer the fuel from the fuel tank to the fuel pump and from the fuel pump to the carburetor. A short, flexible hose is placed at the fuel pump end of the line from the tank side. This hose absorbs the vibration in the line between the engine and the frame. Fuel filters or screen are used to remove the dirt and water from the gasoline. The fuel lines are located so as to keep vapor lock to a minimum. The main feed line is located on the outside of the frame side rail, away from the exhaust system. Vapor lock is caused when the fuel in the lines or pump becomes heated and the fuel begins to vaporize. This vaporization causes tiny air bubbles, and the fuel flow is reduced or stopped.

The fuel pump is operated by a cam on the camshaft. When the highest part of the cam comes around, it pushes on the fuel-pump rocker arm and, through a linkage connected to a diaphragm, a vacuum is created in the fuel chamber. Gasoline is forced through an inlet valve into the fuel chamber by the atmospheric pressure on the fuel in the fuel tank. As the low part of the cam comes up, a spring forces the diaphragm up—creating a pressure on the gasoline in the chamber. This closes the intake valve and opens the outlet valve, forcing the gasoline up to the carburetor.

The carburetor produces a vaporized mixture of fuel and air in the proper proportions so that the engine will run at various loads and speeds. The average mixture is one part gasoline to 15 parts of air (by weight). By volume, that amounts to about 10,000 gallons of air to 1 gallon of gasoline. If you thought the carburetor was a hard-working piece of equipment, you were right. Think about the importance of the air supply.

The intake manifold directs the fuel-air mixture from the carburetor to the cylinders, and the exhaust manifold collects the burnt gases and forces them through the muffler system and away from the passenger compartment. The importance of the heat-control valve is covered in Chapter 10.

FUEL TANK SERVICE

To remove the fuel tank for cleaning, repair, or replacement, first drain the fuel tank and put the fuel away in containers stored in a safe area. You don't want someone to come along with a cigarette and put an end to your work. Disconnect the fuel pipe and the gauge wire from the tank unit. Remove the bolt holding the filler neck to the body. You will have to work through the filler door. Pull the filler neck out of the tank. Remove the tank support straps and remove the tank.

If the tank is leaking because of rust spots, it's better to install a replacement tank. Usually one rust spot begets another and you will be continually fixing the tank. To repair a crack or a seam leak, the tank should be brazed. Don't attempt this; have the tank repaired by the experts. After repairs are made you might consider using a tank sealing compound. Chemical welding might work for cracks but it will not help in rust.

To clean a tank, you should have it steam cleaned so that all the varnish is removed. If the vehicle has not run for some time, it's a good idea to clean out the lines. Disconnect the line at the fuel pump and blow air toward the rear. Don't force the crud up the line. Clean the screen by removing the tank unit and blowing the screen clean. Replace the unit. If the gasket is torn use a silicone fix. If the fuel line is dented or kinked, remove the damaged section and replace it with 5/16-inch-diameter tubing. Use flare nuts and double laps. You could replace the section with a piece of hose. Make sure the connections are tight and use clamps if needed.

This is a good time to check the tank unit if you have had a problem with the fuel gauge not registering properly. It certainly is inconvenient to run out of gas when the fuel gauge reads Full.

Clean away all the dirt that has collected around the tank unit terminal and see that the terminals are tight. High-resistance is a common cause of faulty fuel gauge readings. The gauge consists of the dash unit mounted in the instrument panel and the tank unit that is installed in the fuel tank. These two units are connected by a single wire, and each unit is grounded at its mounting location. See that the wire from the ignition switch to the dash unit is correct. Use a continuity tester to check for current. Check for current at the tank unit with this same tester.

If there is current at both units it might be the tank unit that is causing the problem. Remove it and check that the float arm does not stick. You can use another good tank unit as a tester to check out the dash unit. With the ignition switch off, and the battery ground cable disconnected, remove the brown wire from the dash unit. Connect the good tank unit to the dash unit, using insulated wire and spring clips. Make sure of a good ground connection. Connect the battery cable and turn on the ignition switch. Move the arm of the tester back and forth slowly and observe the dash unit. If the dash unit is fine, the pointer will move "E" to "F" freely. If the pointer doesn't move, or moves only part way, the dash unit is defective and must be replaced.

FUEL PUMP SERVICE

The fuel pumps used on both six-and eight-cylinder models are of the diaphragm type and of similar construction. Both pumps are located on the lower, right front corner of the engine. On the six-cylinder models, an eccentric on the camshaft actuates the pump rocker arm. The eight-cylinder pump rocker arm is actuated by a push rod. Pump pressure on the six-cylinder model is 3 1/2 to 4 1/2 psi and on the eight-cylinder model it is 4 to 5 1/4 psi.

The fuel pump consists of a body, rocker arm and link assembly, fuel diaphragm, oil seal assembly, diaphragm spring, cover and inlet and outlet valves. Before testing the pump, tighten the mounting bolts, cover-to-body screws, and inlet and outlet connections. Check for bends or kinks in the lines that would reduce fuel flow. Make sure there is no gasoline in the tank.

You will need a fuel pump tester that will measure both pressure and volume. You will also need a container that has a minimum capacity of one pint. A plastic bottle is fine. Be careful about fire when making these tests.

Disconnect the fuel pipe at the carburetor and place the bottle at the end of the pipe. You might encounter some gasoline spray when you undo the pipe. Start the engine and run it at 1000 rpm. The pump should deliver 1 pint of fuel in 45 seconds. If no gasoline comes out of the pipe, then either the pipe is plugged or the pump is inoperative. Bubbles in the fuel indicate an air leak in the inlet line. Disconnect the lines and blow air through them, and recheck.

Attach the fuel pump tester to the end of the fuel pipe and run the engine at 500 rpm. Note reading on pressure gauge. If the pump is operating properly, the pressure will be 3 1/2 to 4 1/2 on six-cylinder models, and 4 to 5 1/4 on eight-cylinder models.

Before you replace the pump because of low volume or low pressure, you should do a pump inlet vacuum test. Disconnect the fuel line from the pump inlet flex line. Attach the fuel pump gauge to the flex line. Leave the fuel pipe in the bottle to prevent gasoline from spraying about. Start the engine and let it idle until the highest vacuum reading is reached.

Stop the engine. The reading should be at about 10 inches and remain steady. If the reading is below 10 inches or if it falls off rapidly when the engine is stopped, remove the flex line and attach the gauge directly to the fuel pump. If a low reading or a fall-off continues, the pump is defective. If the reading is now 10 inches or more, the flex line was defective.

You can test the entire line by disconnecting it at the fuel tank and attaching the gauge. If the reading drops off or is low, an air leak in the line is the problem. Remember this is a vacuum reading and not a pressure reading. Keep in mind that you are running on the gasoline in the carburetor. Make these tests as quickly as possible.

When the pump pressure test is below specifications it might be due to wear, a ruptured diaphragm, or dirty valves in the fuel pump cover. If pressure is above, it might be due to a tight diaphragm, fuel between the diaphragm material, or a seized rocker arm and linkage. The pump will have to be repaired or replaced.

Repair kits are available that contain seals, valves, and a

neoprene diaphragm. Pumps that are of the sealed type can be exchanged for rebuilt units. Compare the costs, perhaps all the pump requires is cleaning. This is common if the valves are gummy due to old gasoline in cars that are not run often.

To remove the pump, disconnect the line connections and then the flange fasteners. Remove the pump. Remove the push rod on eight-cylinder models. Cap the line openings and cover the block opening. Wash the exterior of the pump to remove all dirt and grease. If you exchange the pump, check for a pump number on the casting. Get exactly the same pump with exactly the same rocker arm. You can turn the cover so that the lines fit but you can't do much with the rocker arm.

FUEL PUMP DISASSEMBLY

Remove the fuel pulsator diaphragm plate and the diaphragm from fuel cover. Mark edges of fuel cover and body flange with a prick-punch indentation. Remove cover screws and lock washers. Separate the fuel cover from the body by jarring the cover loose with a light blow. On eight-cylinder models, unhook the fuel diaphragm from the fuel link by compressing the spring lightly and moving the entire diaphragm assembly away from rocker arm. Slide the eye of the pull rod off the hook of the lever. Lift the diaphragm spring and lower the spring seat out of the body.

On six-cylinder models, raise the fuel pump link with a screwdriver. Unhook the diaphragm from the link by pressing down and away from the rocker arm side. Remove the oil seal and retainer from diaphragm. Clean and rise all metal parts in solvent. Blow out all the passages with air hose. Inspect the pump body and fuel cover for cracks, breakage, and distorted flanges. Examine all screw holes for stripped or crossed threads.

On eight-cylinder models, inspect the diaphragm pull rod oil seal in the pump body. See that you can obtain a new seal before you pry this old one out. Inspect the rocker arm and link for excessive wear and for a loose hinge pin. Check the diaphragm for cracks and creases. See if the valves are stuck in either the open or closed position. If they are stuck they will have to be replaced.

Obtain a repair kit and discard all the old parts that are duplicated therein. To remove the old valves, clear out the staked material with a sharp scriber and pull the valve out with a hook shaped tool. Install the new valves by placing the gasket in the recess and pressing the valve in. The outlet valve cage must face the bottom of the cover, and the inlet valve cage must face opposite. Stake the valve in place. On six-cylinder models, assemble the oil-seal spring (upper retainer, two leather seals, and lower retainer with convex side out). This seals the fuel pump from any oil that might come up from the crankcase.

Replace the diaphragm with a neoprene material and not gasoline-resistant cloth. This cloth has a tendency to leak. It may be original but your asking for problems. Raise the fuel pump link with a screwdriver, install the diaphragm spring, and hook the diaphragm pull rod over the end of the link.

On eight-cylinder models, if push rod seal was removed, assemble new seal in retainer so that the seal fits inside retainer. Start assembly straight in (bore in body with retainer facing out). Use a 7/8-inch-diameter socket to press the retainer down. Stake in place. Set the diaphragm spring on the staked-in seal and place the retainer on top of the spring. Push the diaphragm pull rod through retainer, spring, and oil seal. The flat of the pull rod must be at right angles to fuel link. Compress the diaphragm spring lightly, invert the pump body, and hook the diaphragm pull rod to the fuel link. Place a new pulsator diaphragm over fuel-cover opening, install a plate, and retain with the screw and fiber washer. Install a cover on the body (lining up the prick punch idents). Push on the rocker arm until diaphragm is flat across the body flange. Install cover screws and lock washers loosely. Push the rocker arm through its full stroke to flex the diaphragm and hold in that position while tightening cover screws securely.

If the rocker arm push rod was removed for inspection, use some heavy grease to hold it in place while replacing the adapter gasket and adapter. Use gasket cement. Install the fuel pump gasket and coat it with gasket cement. Mount the pump. If there is some resistance for it to go all the way against the engine block, the camshaft eccentric is probably in position. Rotate the engine slightly with the starter to move the camshaft. Screw the bolts in while holding the pump against the block. Connect the lines by screwing them in with your fingers first. This prevents cross threading. Check the pump and fittings for leaks (after the engine is running).

AIR CLEANER

The air cleaner is mounted on the carburetor air horn and does three jobs: it removes dust and dirt from the air that is taken into the carburetor and engine; it muffles the noise of the air rushing into the engine; and it acts as a flame arrester—should the engine backfire through the carburetor.

Two types of air cleaners are used: the oil-wetted type and the oil bath type. The oil-wetted type is also known as the standard air cleaner. It has an element consisting of a metallic gauze filter that is saturated in engine oil. As the air filters through this element, dirt and dust are deposited on the oily surfaces of the gauze. This gauze also quenches any flame that might be caused by engine back fire.

To service this type of air cleaner, remove the cover wing nut, cover, and filter element. Wash the filter element in solvent or kerosene. Let the element dry and then pour or squirt engine oil over the mesh, draining off the excess. Install the element and cover. Don't overtighten the wing nut; it might affect choke operation.

The oil bath type or heavy-duty air cleaner is used on vehicles that are operating in dusty conditions. This air cleaner is interchangeable with the standard air cleaner and will not affect power or economy in any manner. Air entering this cleaner must reverse its direction directly above the oil level in the cleaner body. Because the dirt and dust particles do not change direction as easily as the air, they fall into the oil instead of continuing into the air stream. Oil is carried up into the gauze with the air and it is caught on the gauze, where it removes any remaining dust. The predetermined amount of oil being carried into the gauze washes the gauze and carries the dirt back to the cleaner body.

Oil bath elements should be cleaned every 1000 miles by washing the filter element in solvent or kerosene. Do not wash in gasoline; it affects the carburetor. In dusty conditions service more often. Check the cleaner body for dirt and service as required. Wash and wipe dry. Fill with one pint of SAE 50 engine oil during the summer season. Use SAE 20 for temperatures below freezing. Assemble the filter and cover the assembly to the body of cleaner. Install a cover wing nut. Install a cleaner, making sure it fits tight and is set down securely. Tighten the clamp and mounting screws. The air supply should now be fine and you can have a look at the carburetor.

CARBURETOR CARE

The carburetor's purpose is to produce a combustible, vaporized mixture of gasoline and air that is distributed to the engine cylinders. Lets have a look at the various systems and then the adjustments. If you understand the systems, you will be able to solve most of the carburetor problems.

The float system controls the amount of gasoline in the fuel bowl. A float bowl acts as a reservoir to hold a supply of gasoline throughout the entire range of engine performance. A float rises in the reservoir as the gasoline enters from the fuel pump and causes a needle valve to be pushed against a needle valve seat. This closes the inlet, preventing further delivery of gasoline into the float bowl until some of the gasoline is used up. When this happens, the float drops and releases the needle valve from the seat so that more gasoline can be delivered. This is a continual process (with the needle opening and closing and just balancing the amount of fuel being used).

The idling system controls the amount of gasoline to the engine during idle or low-speed operation. Gasoline flows from below the throttle plate to the engine. An adjusting screw at the base of the carburetor controls the amount of fuel mixture flowing to the engine. Turning the screw inward causes less fuel mixture to flow. The result is a "lean mix" (more air to gasoline). Turning it outward "richens" the mixture (giving more fuel than air). There is also an idle speed adjusting screw that controls throttle opening that in turn controls the engine speed.

As the engine speed increases, because of more throttle opening, gasoline starts to flow from the fuel bowl, through the metering jet, and out the main nozzle into the throat of the carburetor where it atomizes into the air flow. The idle-passage fuel mixture is slowly reduced as the main nozzle delivers more gasoline. These two systems interact and produce a smooth fuel-air flow at all engine speeds.

A power system provides additional fuel for heavy load and high-speed engine requirements. When the engine vacuum is high, the power valve is closed. When engine vacuum drops with load, the power valve starts to open. This allows additional fuel to enter the high-speed circuit to enrich the mixture according to engine load and speed.

When the throttle is opened suddenly, for quick acceleration, a plunger type of pump, operated by the throttle linkage, discharges a stream of gasoline into the air stream. Without this extra gasoline, there would be a drop of power on acceleration because the gasoline flow through the high-speed circuit decreases.

The high-speed system continues to cut in more and more while the low-speed system continues to cut out until the vehicle reaches a speed of 30 mph. At this point, the high-speed system is carrying the entire load and the idle system ceases operation. You need one more circuit to complete the carburetor and this is the one designed for cold-weather starting. In cold weather, the gasoline vapors condense when they contact the cold engine parts and little gasoline is left for starting. The choke provides the rich mixture so that a combustible mixture can be drawn into the cylinder. A choke valve at the top of the carburetor controls the amount of air entering the air horn.

The choke system includes a thermostatic coil, housing choke piston, choke valve and fast-idle cam and linkage. It is controlled by a combination of intake manifold vacuum, air velocity against the offset choke valve, atmospheric temperature, and exhaust manifold heat. When the engine starts, the choke must be partially opened to prevent flooding. As the engine starts to warm up, a leaner mixture is required. A fast-idle mechanism, which is in linkage with the choke lever, is an aid to more efficient cold weather warm-up, ensuring correct idle speeds during warm-up period.

CARE, MAINTENANCE, AND CARBURETOR ADJUSTMENTS

Dirt, gum, and varnish are the main enemies of carburetors. Once they get into the passages, you will have problems with rough idle, poor performance, and lack of response to mixture adjustments. Before you blame the carburetor, make sure the rest of the engine is in good working order. Ignition and compression come a long way before fuel.

If you can smell gasoline when the hood is up or when sitting in the passenger compartment, you better have a look at what is happening. Fuel overflow is a definite fire hazard. The fuel inlet line might need tightening but usually it's a problem with the needle and seat. It is possible that a speck of dirt is stuck between the needle and the seat permitting the gasoline to keep flowing. Tap the top of the carburetor where the inlet line enters and see if you can dislodge the problem. If the "flooding" continues, check the fuel pump pressure; it may be excessive. If the pressure is fine, it is a float level problem. The top of the carburetor has to be removed.

Disconnect the fuel line and the line from the choke to the heat source. Remove the cover-attaching screws. Disconnect the fast-idle linkage. Lift the cover straight up and place the assembly on its top. Pull out the float-hinge pin, float(s), and float needle. Examine this needle for any wear on its sealing and seating surface. If there is any sign of a groove, then the needle will have to be replaced. Replacement needles come with a seat so the entire assembly is replaced.

Fuel level in the bowl should be checked as too low a level will cause a lean mixture showing up as stalling and poor performance. A high level causes flooding, hard starting, and rough idle. With the float(s) in place and the cover gasket in position, measure the height of the float(s). On Rochester Model "BC" 1 9/32″ there are two floats. Make sure they are level. Bend the float arms to obtain the proper height. See that the floats are parallel in order to prevent rubbing the inner sides of the float bowl. Hold the cover assembly right side up and measure the float drop. The bottom of the float should be 1 3/4 of an inch below the gasket. Bend the float tang at the rear of the float assembly to obtain the proper drop.

For Rochester Model 2GC carburetors used on eight-cylinder models, the float height is 1 1/4 of an inch and the float drop is 1 29/32 of an inch. The gasket must be in place.

Check the fuel in the bowl for contamination by dirt, water, gum, or metal particles. This is trying to tell you something so pay attention. This garbage ends up in the carburetor. If you are buying gasoline at a discount, think about it. Inspect the gasket surfaces between the body and the air horn. Rough idle may be due to air leakage (causing a lean mixture).

Check the accelerator pump plunger; it should not be damaged in any way. If it is damaged, you will experience the familiar "flat spot" when you step on the throttle. Place the cover on the bowl and securely tighten the attaching screws. Reconnect the choke lever and choke tube. Connect the fuel line. Start the engine and bring it up to operating temperature. Make sure the choke is open. If it is not open then the engine is running on a rich mixture. This should be evident by the black smoke coming out of the exhaust.

If you've been having trouble with hard starting, it might well be the choke and not the carburetor that is causing the problem. An improperly adjusted choke or one that is sticking will cause just about the worst gas mileage imaginable. Keep in mind that the excess fuel is not being burned. It is going down into the oil pan, causing oil dilution, and you can see the importance of having the choke set right. Don't get in a panic about changing over to a manual choke. Adjust or clean the automatic choke. It does a good job but it has to work right.

The normal setting of the choke is such that the scribed index mark on the choke cover is in line with the long mark on the choke-housing casting. Check where the mark on the choke cover is now and make a note of it. The choke valve must be free in the housing. That is the choke shaft must not be seized in the choke shaft bushings. Use carbon solvent cleaner in the spray can on these bearings.

Remove the choke heat tube and see that it is clean inside. Use more carbon spray. Remove the choke cover and the thermostatic coil assembly from the choke housing. Clean with carbon solvent and be careful not to distort the spring. Inspect the inside of the thermostat spring housing and the vacuum piston. It must move freely. If it is stuck, try some more carbon solvent or you will have to remove the piston for cleaning.

Reassemble all the choke parts. Make certain the thermostatic spring is positioned correctly and that it engages the choke shaft lever. Align the housing and install the three retainers and screws to the choke housing and tighten securely. Check the choke valve for free movement.

The choke valve should be slightly closed at room temperature (85 °F). Bring the engine up to normal operating temperature. The choke valve should be wide open. If it isn't and black smoke is coming out of the exhaust pipe, then loosen the cover screws and turn back just enough to open the choke valve. Some covers are marked Lean or Rich. This helps to indicate which way they should be moved. Hard starting, sputtering, spitting, and coughing during warm-up might indicate the need for a richer setting. Engine loping or black exhaust smoke might indicate a leaner setting.

There are two more adjustments you can make but it's difficult to do a good job without the use of a vacuum gauge and a tachometer. The tachometer hooks up to the distributer side of the coil and a "bright" ground connection. The vacuum gauge hooks up to the intake manifold. Check the intake manifold torque and the carburetor-to-intake manifold torque to make sure there is no air leak. See that the manifold heat control valve is free of binding. Spray with carbon solvent.

With the engine at operating temperature, see that choke is entirely open and the linkage is free. See that the throttle stop screw is against the low step on the fast idle cam. Set the parking brake tight or have a helper on the brake pedal. On Powerglide models, place the selector lever in Drive (D) range and on standard in

Neutral. Screw the throttle stop screw in or out to obtain an idling speed of 425 rpm on Powerglide models and 475 rpm on standard models.

The idle mixture adjustment should be set to give peak vacuum and rpm indications on tachometer and vacuum gauges. Missing is an indication of too lean a mixture and loping is too rich. If you can turn the idle adjusting screw all the way in and the engine continues to run, then the carburetor needs cleaning. The correct adjustment is 2 1/2 turns, and then either way from this position until best idle is reached. Re-adjust idle speed and re-check the mixture. On eight-cylinder models, there are two idle mixture screws. Adjust one for the highest rpm and smoothest running. Then adjust the second screw the same way. Go back to the first screw for readjustment and then perhaps the second until the carburetor is properly balanced. The starting point is 1 1/2 turns until the best idle.

To ensure wide open throttle with full accelerator depression, there is a throttle rod adjustment. On six-cylinder models, the adjustment is made at the bell crank on the left side of the engine block. With the accelerator pedal fully depressed and the carburetor throttle fully opened, the swivel should be adjusted on the control rod for free entry into the bell crank. Then turn the swivel two full turns to lengthen the throttle control rod. Assemble the rod to the bell crank. On eight-cylinder models, adjust the threaded swivel at the throttle lever to suit.

REBUILD OR REPLACE THE CARBURETOR

First, make the above adjustments and see if performance improves. If it does ever so slightly, do an on-car carburetor cleaning. Purchase a can of spray cleaner and clean the outside of the carburetor and the related linkages. Now, purchase a carburetor cleaner that you can inject directly into the carburetor. Follow the manufacturers instructions. What are the results?

Well there is but one good way to clean the internal passages and that is to disassemble the carburetor and soak everything in carburetor cleaner. You can do this or you can buy a rebuilt or exchange carburetor. Check the costs of a carburetor overhaul kit and your ability against a factory or jobber rebuilt. Do what you are comfortable with, but a carburetor overhaul is not difficult if you take your time and follow the instructions.

Order a complete overhaul kit that comes with gaskets, needle and seat, accelerator pump, various paper templates and a sheet of instructions. There might be a small brass tag under the cover mounting screw that will identify the carburetor. Use this information to order the kit. You will also need about a quart of carburetor cleaner. Ordinary solvent will not loosen up old gum, varnish, and dirt or rust.

The following procedure applies to the Rochester Model 2GC Carburetor, which is standard equipment on all eight-cylinder models. If you are overhauling the Rochester Model BC used on six-cylinder models, it's a bit easier but follow the overhaul instructions.

Remove the air cleaner wing nut, air cleaner, gasket, and stud. Disconnect vacuum line, spark control line, fuel line, and choke-heat tube. Disconnect the throttle rod at carburetor and remove throttle return spring. On Powerglide models, disconnect the transmission control rod from the throttle lever. Remove the four nuts and washers that hold the carburetor to the manifold and lift the carburetor straight up and out. Keep it upright as you want to examine the contents of the fuel bowl.

Clamp the largest taper punch that you have (vertically) in the vise jaws. Set one of the mounting holes in the throttle body on the punch. You can disassemble the carburetor on your work bench but the above helps to keep the bowl level. Remove the retaining screw at the end of the choke shaft and carefully pry off choke trip lever, and the fast-idle link and lever. The lever can be removed from the link by turning until the slot in the lever will pass over the tang on the link.

Remove the three choke-cover attaching screws and retainers, then remove choke cover and thermostatic spring. Do not remove the choke valve from the choke shaft lever. Disconnect the pump link from the throttle lever by removing the retainer. The link can be removed completely by rotating it until it clears pump lever. Remove eight cover screws and lift the cover (air horn) from the float bowl. Check the fuel in the bowl for contamination by dirt, water, gum or other foreign material. A magnet moved through the fuel will pick up any iron oxide dust that might have caused needle and seat leakage. Safely dispose of the remaining fuel.

Place the upended cover on the bench and remove the float hinge pin and lift float assembly from the cover. Shake the float; anything in it? Remove the float needle seat with a proper-sized screwdriver and remove the filter screen. Remove the power piston by depressing shaft and allowing the spring to snap. This forces piston from the casting. Remove retainer on pump plunger shaft, and remove the plunger assembly from the pump arm. The pump lever and shaft can be removed by loosening set screw on inner arm and removing outer lever and shaft. Remove the cover gasket.

Remove the pump plunger return spring from the pump well. Remove the main metering jets and power valve. Use a screwdriver that fits properly. Remove the three screws that hold the venturi cluster to the main body, and remove the cluster and gasket. Remove the pump discharge spring retainer. Use a pair of long needle-nose pliers. Then remove the spring and check the ball. Up-end the main body and remove the throttle-body attaching screws. The throttle body and gasket can now be removed. Remove the idle-adjusting screws and springs from throttle body. Do not remove throttle shaft.

Clean the carburetor castings and metal parts in the carburetor cleaner. Do not soak the pump plunger or the thermostatic spring. This cleaner is caustic so keep exposed skin and eyes away. About 30 minutes of soaking should do the job. Neutralize as per instructions (water or solvent rinse). Remove the pieces one by one and blow all the passages with compressed air. You must get all the crud out. Do not clean any of the passages or jets with wires or drills.

Check all parts for wear and replace as necessary. Use all the new parts in the kit. Set the float level as per instructions. Back the idle mixture screws out 1 1/2 turns as a temporary idle adjustment. Set the pump adjustment and bend the accelerator pump rod if necessary. This measurement is taken with the throttle valves fully closed from the top of the pump housing to the top of the pump rod. The distance is 57/64 of an inch. The fast-idle adjustment is measured when the idle-adjusting screw is on the next to highest step of the fast idle cam. Use a 3/32-inch drill bit and see if it slides between the upper edge of the choke valve and the bore of the carburetor air horn. If necessary, bend the choke rod until the required clearance is obtained. Hold the drill bit vertically when measuring.

To prevent the possibility of the choke closing during heavy load or acceleration, there is an adjustment on the tang of the throttle lever (which is the unloader adjustment). Place the throttle in wide open position. See if a 3/8-inch drill bit will slide freely between the upper edge of the choke valve and the bore of the carburetor air horn. If an adjustment is necessary, bend the tang of the throttle lever to obtain the clearance.

Set the choke valve so that the index mark on the cover is aligned with the long mark on the choke housing. Remember, this is the starting point and you can make corrections. Think about how the thermostat in your home works and you will get a better idea of what the automatic choke is trying to do.

Use a new gasket on the intake manifold and install the carburetor over the manifold studs. Start the spark control pipe fitting into the carburetor fitting. Install and tighten the four washers and nuts (copper washers on the two front studs). Tighten the spark control pipe fitting and connect and tighten the choke heat tube, vacuum line and fuel line. Install the throttle rod and throttle return spring. On Powerglide models, install transmission control rod.

Check the intake-manifold-to-cylinder-head torque (25 to 35 foot pounds). You will have to prime the carburetor unless the fuel bowl was filled previously. Don't prime and attempt to start the engine at the same time. A backfire could result. Inspect the manifold heat control valve for freedom of action and correct spring tension. Start the engine and allow time for warm-up. Adjust idle speed and mixture.

Install the air cleaner gasket and stud. Service the air cleaner and install it on the carburetor. Tighten the wing nut securely. There should be no change in idle speed. A dirty air cleaner will produce a rich, fuel-wasting mixture.

The carburetor is sometimes blamed for many problems it does not cause, but it can waste gasoline if it is not properly adjusted. If you are interested in fuel economy, check that the engine has proper compression, accurate ignition timing, good spark plugs, and use a light foot on the gas pedal.

Chapter 6

Steering and Suspension

The more you drive the more comfortable you become with your car, and the more forgiving you both become. Try to remember that you both need some care, maintenance, and adjustments. Listen to what your car is trying to tell you and look after the symptoms before they lead to "major surgery." Hard steering can be caused by lack of lubrication. Road wander can be caused by underinflated tires. Take a minute for a visual check every morning, and in only one week you'll both be feeling better. The front suspension has more moving parts than the rear suspension. Therefore, it needs more attention.

With the car on a level surface, check for sags (front, rear, left, and right sides). The engine weight that sits on the front coil springs will in time cause these springs to sag. A broken leaf in the rear spring will cause a similar problem. If you are the driver and sole occupant of this car, then the left hand front may have a slight sag.

Inspect the tires; they should be of similar size and properly inflated. The wear pattern should be across the tread. Check the shock absorbers for proper dampening action by bouncing the car up and down at each corner and letting go at the down stroke. The car should not rebound more than once. Examine the shock absorbers for leaks. There shouldn't be any. Lift the hood and examine the gearbox to frame fasteners. They must be tight. Check for free play in the steering wheel; about 1 inch is acceptable. If it takes more than this to move the front wheels, there is wear in the steering system.

Raise the front of the car on stands, placing the stands under the lower suspension arm at the outer end. With the wheels hanging free, grab one of the tires at the 3 and 9 o'clock position and move it in and out. Do the same at the other wheel.

Grab both wheels at the front and push in and out. There should be very little looseness. If there is the tie rods, relay rod, or idler arm is loose. Grab the tire at the 6 and 12 o'clock position and move it in and out. Looseness here could be the lower ball joint or the wheel bearings. Put a bar under the tire and the ground and move it up and down. Check for looseness at the upper ball joint. To eliminate the front wheel bearings on this test, remove the hub cap and the dust cap. Remove the cotter pin and tighten the spindle nut. Check where the movement is apparent (ball joints or front wheel bearings). You can save a number of dollars by replacing the worn parts yourself and then having a wheel alignment check done. Use safe working procedures to prevent injury. You will need very few special tools (a fish scale at the most). Steam clean the front end and the related parts.

STEERING GEAR SERVICE

There are two adjustments that can be made to the steering gear if there is excessive play in the steering wheel and if the rest of the steering linkage is satisfactory. These are the adjustment of the worm bearings and the adjustment of the sector shaft. Disconnect the steering relay rod from the pitman arm, which is located at the end of the sector shaft. There are quite a number of pieces here so note the order.

At the top of the gear box, you will notice an adjuster screw and a lock nut. Loosen the lock nut and turn the screw out a few turns to remove the pressure on the worm bearings. Turn the steering wheel gently in one direction until stopped by gear, and then back away about one turn.

Attach the fish scale to the spoke at the outer rim of the steering wheel and pull at right angles to the spoke. The reading should be between 3/8 and 5/8 pounds. If not, the worm bearings need adjustment. Loosen the worm bearing adjuster lock nut and turn the worm bearing adjuster in until snug. Check with the scale and readjust as required. Tighten the lock nut and recheck. If the movement of the wheel seems rough or uneven, there is probably worm-bearing damage, and the assembly must be removed for repair. Make sure that this adjustment is checked when the gear is in an off-center position. In other words not in straight-ahead position.

To adjust the sector shaft, locate the straight-ahead position by turning the steering wheel from the extreme left to the extreme right, and then halfway back. Do not bang against the stops or internal damage might result. Turn the adjuster screw in until all play is removed. Tighten the lock nut. Move the steering wheel off the center position. Attach the scale to the outer rim and pull through the center position. This reading should be between 7/8 and 1 7/8 pounds. Readjust to obtain the necessary pull. Reassemble the steering relay rod to the pitman arm. Check that nothing fell out of the relay rod. The spring, plug and ball seat fit on either side of the pitman arm. Then the end plug. Tighten to remove all end play and back off three-quarters of a turn or enough to insert a new cotter pin.

If you do much long-distance driving, it's nice to have the wheel spokes in horizontal position when the front wheels are in the straight-ahead position. Comfortable hand position makes for effortless control. With the wheels in the straight-ahead position, remove the horn button and check the location of the position mark on the end of the wormshaft. It should be at 12 o'clock and aligned with the mark on the steering wheel. If not, remove the steering wheel and position it properly. Use a steering-wheel puller to prevent damage to the wormshaft. The steering wheel hub is threaded, and it's an easy matter to make a puller that will do this job.

If the gear has been moved off the high point when setting the wheels in straight-ahead position, the tie rod adjuster sleeves will have to be moved. Make this correction carefully because it affects the toe-in setting. Loosen the adjusting sleeve clamps on both left- and right-hand tie-rods. Then turn each adjusting sleeve an equal amount in the *same* direction to bring the gear back on center. Tighten sleeve clamps and locate bolts below tie-rods to avoid frame interference.

The tie-rod ends are self-adjustable for wear but will have to be replaced when excessive up-and-down motion or endplay exists. The tie-rod ends are provided with right- and left-hand threads; buy the proper ones. If both ends of the tie-rod are worn, remove the entire tie-rod so you can work on the upside instead of the downside. Remove the cotter pin and the fastening nut. Use a heavy hammer as a backing and strike the connecting part with another hammer. Pull out the tie-rod. Remove the other end if necessary. Measure the new part against the old part so that it can be installed to the same distance and keep the toe-in the same. Toe-in should be 1/16 of an inch to 3/32 of an inch. Install the ball-stud nut, tighten securely, and install a new cotter pin. Lubricate. If the relay rod or idler arm need new bushings, attend to them now. Pitman arm ball seats are available if inspection shows need.

REPLACING BALL JOINTS

Now is the time to replace the front springs if inspection shows they are becoming weak and starting to sag and bottom. You should also check the cross shafts and bushings on the upper and lower control arms to see if they are worn. This is a service that requires press work, but if you take it apart you will save considerable dollars. See that the parts are available before you get this far with the disassembly. You will need a hydraulic jack and a short piece of chain to prevent the coil spring from flying out when you lower the control arm. This can be dangerous work because the compressed coil spring contains plenty of energy to knock your teeth out and more. Remember PPPP—proper procedures prevent problems.

Raise the front end of the car and place stands beneath the frame-side rails. Remove tire and wheel assembly. Remove the front shock absorber. Thread a piece of chain through the coil spring and around the lower control arm. Leave enough slack in the chain so that the spring could be removed when the pressure is off. Bolt the ends of the chain together—tight. Place the hydraulic jack under the outer end of the lower control arm and raise it just enough to place some tension on the coil spring. Remove the lower ball-joint cotter pin and nut. Loosen the stud by hammering on the side of the steering knuckle joint. Use a heavy hammer against the other side as a support piece. When the joint is loose, lower the jack slowly until the lower control arm is free. If you are replacing the upper ball joint, loosen it in a similar way. Tie the drum and knuckle assembly out of the way; don't leave it hanging on the brake hose.

Check for excessive movement in the upper and lower control arms by moving the assemblies sideways. Decide on whether the coil springs need replacing or retempering. Check the price of retempering against that of new springs. If the cross shafts or bushings need replacement, the control arms will have to be removed from the cross member. Mark the number and location of the shims behind the upper cross shaft. They adjust the wheel alignment. Service as required. To replace the ball joints on original equipment, it is necessary to drill and chisel the heads off the retaining rivets and drive the rivets out. Replacement ball joints are bolted in. Follow manufacturer's specifications for torquing and installation.

If cross shafts have been serviced, leave the capscrews at the ends loose for now. Bolt cross shafts in place and replace shims as noted. Position the spring in the lower control arm, chain in place, and raise the lower control arm with the hydraulic jack. Install the drum and knuckle assembly to the ball joint studs, you might have to pull down the top one. Install the nuts, tighten securely, and install cotter pins. Use extreme caution. Remove the chain. Replace the shock absorber if it is a "leaker" or previous tests show it isn't absorbing spring bounce. Lubricate the ball joints. Service the other side of the vehicle. Install wheels and lower the vehicle to the floor. Bounce the car to centralize the bushings in the cross shafts and now tighten the capscrews. Have the wheel alignment checked at a shop that specializes in this service.

SPRINGS AND SHOCK ABSORBERS

The independent front suspension uses coil springs and direct-acting shock absorbers to provide ride control. The shock absorbers are mounted between the spring seats on the lower control arm and the spring pockets in the front cross member. Rubber stops are mounted on the lower control arm and the frame cross members to prevent metal-to-metal contact when the front suspension is operating under very severe road conditions. If these stops are broken up under normal road conditions, it means the springs and shock absorbers are in need of replacement.

The rear springs are semi-elliptic (consisting of four leaves each). The top leaf is shot-peened for long life and the second and third leaves have wax-impregnated fabric inserts at the leaf ends (for quiet operation). Rubber bushings are used to mount the spring to the frame. The springs are fastened to the axle housing by U-bolts. The rear shock absorbers are of the nonadjustable, direct-acting type. They are fastened to the rear body floor and to the bottom of the rear-spring, anchor-bolt plate.

To replace the front coil springs, the ball joints must be disconnected from the steering knuckle. This information is covered in a preceding section. The job is time-consuming but it can be done to save money. Coil springs can be retempered so check cost of new against old. Remember, as in all operations, safety comes first. If it feels right do it. If not leave it. To replace a sagging rear spring or one with a broken leaf is also time-consuming but it's a job you can do if you prefer.

Raise the vehicle and place jack stands under the frame side rails. Remove the wheel on the side you are working on to give yourself some room. Support the rear axle housing with a jack stand. Soak the "U" bolt nuts with penetrating oil. Check the ends of the "U" bolts for damage and use a file to taper the ends. If the shock absorber is fine, leave it on the anchor plate. Remove the "U" bolt nuts.

If necessary, adjust the jack stand under the rear axle housing so that the spring hangs free. Remove the two self-locking nuts from the shackle pins and pull the shackle plate off the shackle pins. Rotate the spring eye around the hanger eye to position the spring eye and provide clearance for removing the shackle. This method provides an opportunity to examine the rubber bushings for wear or deterioration. Examine the shackle plates and pins for cracks or defects. Alternately, you can remove the front bushing nut and bolt first, and then pull the spring off the rear shackle pin.

To replace a broken leaf or put in an extra leaf to correct sagging, the spring assembly must be taken apart. Place the spring in a vise and remove the spring clips by bending tabs of lower half of clip. Slowly open the vise. If all the leaves come loose, the problem is a broken center bolt. Place the spring in the vise jaws, compressing the leaves at the center, next to the center bolt. Remove the center bolt nut. Open the vise slowly to let the spring expand.

Examine all the parts. If one leaf is broken, examine the others for any signs of cracks. Check the spring leaf inserts for wear and replace if necessary. The inserts are removed by driving the rivets out. Check the front eye bushing for wear, damage, or deterioration caused by lubrication. Check the rear shackle bushings for similar problems (there are four bushings here). Replace the spring clips if the tops are broken off. Replace the center bolt. Do not use the old one.

To remove the front bushing in the main leaf, use a press and replace the new one the same way. Clean all the leaves and give them some color if you are so inclined. Sagging springs is the problem to correct so check on the price of retempering over the price of a new assembly. You might consider adding an extra leaf just under the main to firm up the spring assembly.

Buying a spring from the auto wreckers will probably get you one just about as good as the one you have. Replace broken leaves with new ones from a spring shop. To assemble the leaves, use the vise or a C-clamp to compress the leaves and the center bolt to align them. Do not use the center bolt as a spring compressor. See that the leaves are in line and securely tighten the center-bolt nut. Cut the rest of the bolt off, leaving about 3/16 of an inch for peening. Assemble the spring clips by bending the tabs of the lower halves over the upper halves. The spring clips align the leaves and should not bind the spring action.

Place the front of the spring in the hanger eye and install the through-bolt and nut. Do not tighten, leave loose for now. Install the bushings on the rear shackle pins—large end facing out—and place the shackle in the hanger eye and the spring eye with the curved side facing up. Rotate the spring to normal position and install the other two bushings. Install shackle outer plate to match the inner plate and with the raised nut facing outward. Install nuts and leave loose. Index the center bolt head with the hole in the spring seat on the axle. Install the U-bolts and the anchor plate. Tighten the nuts evenly and to 90 foot pounds of torque. Replace the wheel and lower the vehicle to the floor. Bounce several times to centralize the bushings and then tighten the front bushing nut 60 to 90 foot pounds torque and the rear shackle nuts to 25 to 30 foot pounds of torque.

To replace the rear shock absorbers, you should raise the car on jack stands under the rear axle housing. However, you can do the job with the weight of the vehicle sitting on the wheels. The upper stem of the rear shock absorber extends through the body floor and into the trunk compartment where the retaining nuts are found. Hold the upper stem from turning and remove the nut, retainer, and grommet. Now slide below and remove the nut, lock washer, and flat washer from the shock absorber anchor bolt. Pull the shock absorber off the anchor bolt and remove the rubber grommets. New ones come with the replacement shock absorber. Useful shock absorber life is about 15,000 to 20,000 miles under average driving conditions.

Read the instructions that come with the new shock absorber so that you will install it properly. The rubber bushing goes in the shock absorber eye and the retainer and grommet slide on the upper stem. Install the flat washer on the anchor bolt, which must be tight in the anchor plate, and push the upper stem through the body floor. Slide the lower end of the shock absorber on the anchor bolt. Install the flat washer and nut and tighten securely.

Now, back into the trunk install the other grommet and retainer on the upper stem which is sticking out through the body floor. The two grommets face each other with the floor between them. Install the retainer nut until it bottoms on the stem shoulder and the grommets begin to expand slightly.

Lock the nut in place or stake the single nut. To replace front shocks, hold the upper stem from turning and remove the upper stem retaining nut, grommet retainer, and grommet. Remove the two bolts holding the lower shock absorber pivot to the lower control arm and pull the shock absorber out the bottom of the spring housing. Install the new grommet retainer and the grommet on the upper stem of the new shock absorber. Push it through the spring housing. Install another grommet and grommet retainer on the upper stem. Run the retainer nut down until the grommets just start to expand. Secure the bottom bolts at the lower control arm. There are many makes of shock absorbers on the market. Pick the kind that will go with the condition of the springs.

SERVICING FRONT WHEEL BEARINGS

The wheel and the hub assembly must be removed to service the front wheel bearings. Support the vehicle on jack stands at the lower control arms. Pry off the wheel cover and remove the wheel nuts. Remove the bearing dust cap. Remove the cotter key and the adjusting nut. Pull the drum forward and the outer bearing and washer will come loose. Put them aside. If the brake shoes drag, loosen them so the drum will come off easier. Don't drag the seal on the spindle threads when you pull the drum off. Pry the seal out; it should be replaced. Wash the parts clean and examine all bearings for chipping, galling, and wear. Check the bearing races for cracks or pitting and for looseness in the hub. If the balls are damaged, replace the race.

To replace the race, use a soft steel drift and tap the cup through the slots in the hub. Move the drift around so the cup comes out evenly. Wipe the recess clean and lubricate the new cup so it goes in easier. Use the press to install the cup or a piece of pipe with the correct outside diameter. In absence of the above, use the soft steel drift. Tap a little at a time and move around the cup to prevent tipping. Make sure the cup is not cocked and that it is fully seated against the shoulder in the hub.

Pack the wheel bearings with wheel-bearing grease and lightly coat the inside of the hub. Place the inner bearing in the hub and install a new seal with the bent lugs facing out.

Wipe off any grease that is on the outside of the seal or hub. Carefully place the hub on the spindle and install the outer bearing by pressing it into the hub. Install spindle washer and spindle nut. Proper adjustment of the wheel bearings is important for safety and long life of the bearings and tires. These bearings should not have any end play and a slight preload is fine.

Tighten the spindle nut while rotating the drum until all the end play is removed. Back off the nut 1 flat or 1/6 turn. Check for end play. Insert the cotter pin in the first, nearest hole—either vertical or horizontal—that lines up with a slot in the adjusting nut. If you are using a torque wrench, tighten the nut to 33 foot pounds and then back it off 1/6 turn.

Spin the wheel to make sure it is free and smooth. The cotter pin must be a tight fit in the spindle hole and the ends must be bent back to prevent rubbing the dust cap. Lightly grease the inside of the dust cap to prevent rusting and put it in place. Replace the wheel and wheel cover and do the other side.

To ensure long tire life and ease of steering, you should have the front end aligned. This is a process of checking and adjusting all the interrelated steering system parts. You should also have the front and rear wheels balanced at the same time. See that the tire tread patterns are the same for either front or rear wheels. Tire size and pressure should be as specified.

Deal with a shop that has good equipment and help that knows how to use it. You have made the checks and replaced the necessary parts. Don't let some one sell you a parts changeover. Be alert to what you have done. The shop will check and adjust caster, camber, and toe-in.

Improper caster angle will not cause tire wear but it will cause wandering and pulling to the right or the left. Camber places the tire-to-road contact area nearer to the point of load. This makes for easier steering. Improper camber will cause unstable steering and tire wear. The caster and camber adjustments are made by the shims between the upper control arm inner support shaft and the support bracket attached to the frame side rail. Both adjustments are checked together and adjusted in one operation. Caster should be 0 degrees plus or minus 1/2 degree, and camber should be 1/2 degree plus or minus 1/2 degree.

Toe-in is the placing of the front tires closer together at the front than at the back. This setting is very important as improper toe-in will wear the tire tread off in record time. It is adjusted by increasing or decreasing the length of the tie rods. Proper toe-in distance is 1/8 to 3/16 of an inch.

Steering axis inclination and toe-out on turns are not adjustable, and require replacement of spindles or steering arms. Do not bend or weld steering system parts. Remember PPPP.

Chapter 7

Brakes

The brakes used on the front and rear of all models are the Duo-Servo, single-anchor type. This means that one brake shoe helps to apply the other when the shoes are forced out against the drum. The shoes are also self-energizing because the free end of the shoe is forced into contact with the drum. Each brake has one wheel cylinder located near the top and just below the anchor pin. The linings are bonded to the shoes instead of being riveted. The front brakes are 2 inches wide and the rear brakes are 1 3/4 inches wide. In each brake assembly, the linings for the front and rear shoes differ in length because, in operation, a greater force is applied to the rear shoes than to the front.

When the brakes are applied, the brake fluid in the brake lines and in the wheel cylinders is compressed and the pistons in the wheel cylinders are forced outward—pushing the brake shoes against the turning drum. Friction between the lining and the drum causes the car to stop. Any increase in brake pedal pressure beyond this point will cause an increase in shoe-to-drum contact. When the pedal is released, the brake shoe retracting springs force the brake fluid to flow backward into the master cylinder.

The system is dependent upon a single piston master cylinder. When it fails, total brake failure is the result. Any foreign matter in the system will cause inefficient brake operation. Air will cause a spongy pedal. Dirt or grease on the brake linings will cause erratic brake operation. Don't trust your life to sloppy work or inferior brake parts. Do complete tests with no short cuts.

A mechanical parking brake is located to the left of the steering column. Through a system of pulleys and levers, a brake cable that is attached to the rear brake shoes can be operated. This causes a brake strut to operate the other shoe and both brake shoes are forced against the rear brake drum, allowing the vehicle to be held when parked.

Each time the brake is applied, a small amount of wear takes place on each brake lining. Therefore, a means of adjusting the shoes is provided for by a single adjusting screw at each brake drum. These brakes are not self-adjusting.

BRAKE INSPECTION

Most brake problems develop gradually, and if you are the only driver of the vehicle you become accustomed to the brake system. Therefore, about once a year you should remove the brake drums and check the linings, drums, and wheel cylinders. Replacing worn linings is a job that can be done in a few hours. Drums that need resurfacing can be machined at a local brake shop. Leaking wheel cylinders can be rebuilt while on the vehicle. Let's do some inspections without removing the brake drums.

The brake pedal free play should be 1/16 of an inch. Before making this adjustment, make sure the brake pedal returns to the fully released position and that the brake retracting spring has not lost its tension. The adjustment is made at the brake pushrod located in front of the brake pedal and under the dash. Check the total pedal travel at about half. It should be firm with no spongy feeling that could indicate air in the system. The pedal should remain at one position and not move slowly to the floor. If it does there is an internal or external brake fluid leak. Check the brake fluid level in the master cylinder. Clean all dirt from the top of the master cylinder and remove the filler plug. The level should be 1/2 an inch from the top. If the brake pedal sinks to the floor without loss of brake fluid, the master cylinder needs rebuilding.

External brake leaks can be at the brake lines or at the wheel cylinders. Dampness indicates a leak. This inspection can be made with the vehicle safely supported on jack stands. Dampness around the brake drum or at the bottom of the backing plate indicates a leaking wheel cylinder. This can be checked by removing the brake drum. Check the bleeder valve; it must be tight in the wheel cylinder. Replacing leaking lines is an exchange job. Check the new one against the old one for size and length. All models use 3/16-inch tubing, except on the brake main cylinder pipe and front cross-over pipe. These are 1/4 of an inch. Buy good-quality, double-layer annealed steel line.

It is more convenient to buy the line with the fittings on it but you can make up your own line if the fittings aren't rounded off. Use a double-lap flare at the ends to produce a strong leak-proof joint. To do this, you will need a flaring tool. Unless you already have one, it's less expensive to buy made-up line.

Anytime you replace the brake hose or the steel lines, you will have to bleed the air out of the system. This is a two-person job but you can do it yourself. It's not that difficult. The air is forced out through the bleeder valves located at the top of each wheel cylinder.

Remove the master cylinder filler plug. If the brake fluid is discolored, this is a good time to replace the fluid. The brake fluid has a tendency to absorb moisture so it's not a bad idea to replace the fluid every two years or so. Some new brake fluids on the market have a silicone base. Use a good-quality brake fluid. Your life depends on it. Bleeding is done on the longest line first. The proper sequence to follow is left rear, right rear, right front, and left front.

Raise the vehicle on stands if you have not already done so. Even if you have only replaced the front hoses or lines, it is a good idea to bleed the entire system. If the brake fluid is fine and you are only adding fluid, then you can do this job by yourself.

Remove the bleeder valve screw from the bleeder valve. Fit a six-point, box-end wrench over the bleeder valve. Slide about an 18-inch piece of tight-fitting vacuum line over the bleeder valve. Place the other end into a jar that has fresh brake fluid in it. The end of the hose must stay in the brake fluid. Loosen the bleeder valve about a 3/4 turn. Check the master cylinder for brake fluid and replace the filler plug. Brake fluid eats paint. Remove any excess fluid immediately. Step on the brake pedal and push it all the way down and then slowly let it up. You might be able to hear the bubbling but it will take about two or three pumps depending on the length of line you replaced to bleed all the air out. If the fluid coming out is dark and dirty, the entire system should have the brake fluid replaced.

Tighten the bleeder valve, remove the hose, and replace the bleeder valve screw. Repeat at the next wheel cylinder. Check the master cylinder after each operation so that it will not run dry.

Otherwise you are going to be bleeding air a long time. Check the brake pedal for firmness. If it is still spongy, there is air in the system and you will have to bleed it out.

The pedal must be firm. On completion see that the brake fluid is at a level about 1/2 inch below the top of the master cylinder, and that the filler plug vent hole is clear.

REPLACING BRAKE LININGS

Raise the vehicle on jack stands and remove the wheels. Then remove the brake drums. The rear drums should slide off the studs. If they stick to the brake shoes, adjust the shoes inward. See that all the tension is removed from the parking brake cable. If the drum sticks to the axle-flange wire, brush the flange and apply some penetrating oil. Do not tap on the drum with a steel hammer; use a soft-face hammer. The front brake drums come off with the hubs. Keep the bearings with the drums. Do not depress the brake pedal when the drums are removed.

Examine the brake linings for useful life. Linings under 1/16 of an inch-thick should be replaced. Oil or grease on the linings indicates the seals are in need of replacement (as are the linings). You can't wash them clean. Brake fluid on the linings has to come from a leaking wheel cylinder and they will have to be serviced. Uneven wear on the same shoe is a sign of weak springs that can be caused by stretching them on disassembly. Use the proper brake spring remover and replacer tool. You don't need brake spring pliers.

Unhook the brake shoe pull back springs from the anchor pin and the brake shoe hold down springs. Spread the shoes to clear wheel cylinder connecting links and remove the shoes from the backing plate. Separate the shoes by removing the adjusting screw and spring. On rear brakes, remove the parking brake lever from the secondary brake shoe. Keep parts for each wheel together. Clean and inspect all parts.

Springs that are discolored or stretched should be replaced. Wash the brake drums and examine for cracking, ridges, and irregular wear such as those that are barrel shaped or tapered. Do not weld a cracked drum; replace it. Take the drums and the shoes to a brake shop and have the drums serviced if necessary. The shoes are usually exchanged for a set that has new linings. Drums that are turned might require oversize (thicker) shoes. If you find that over the miles of braking action the drums cannot be serviced, it will be necessary to replace them.

Clean the backing plate and use fine emery cloth to polish the shoe pads (the raised portion of the backing plate that supports the shoe). Inspect wheel bearings and oil seal at both front and rear wheels. Check all backing-plate attaching bolts to see they are tight. Service or replace the parking-brake cable. Lubricate with Lubriplate. Keep your hands free of oil and grease and avoid touching the linings. On the rear brake, attach the parking brake lever and the bolt to the secondary shoe. Lubricate lightly. Make sure the lever moves freely. Don't forget the spring washer.

Lubricate the threads and the socket end of the adjusting screw. Connect the brake shoes together with the adjusting screw spring so that the star wheel faces the adjustment slot in the backing plate. The primary shoe is the short one and it faces the front of the vehicle.

Attach brake shoes to backing plates using the brake shoe hold-down springs. At the same time, engage shoes with wheel-cylinder connecting links. On rear brakes, connect the cable to the parking brake lever and install a strut between the lever and the primary shoe. The strut antirattle spring must be installed with the spring loop to the rear and with loop inside shoe on the left side and outside shoe on the right side. Replace the guide plate and slide the front brake shoe pull-back spring over the anchor, and then the secondary pull-back spring. Pry shoes away from the backing plate and sparingly lubricate the shoe pads. Recheck, and install the brake drums.

Service the front wheel bearings and adjust. Replace the wheels. To adjust the brake shoes you will need a brake adjusting tool. Make one out of a spare screwdriver by bending the blade end at about a 45-degree angle, 1 inch from the tip.

Remove the cover plates with the brake adjusting tool. If you are adjusting the rear first, loosen the check nut at the brake cable equalizer to remove tension from the brake cable. Expand the brake shoes by moving the star wheel clockwise until a light drag is felt on the brake drum. Apply the brake pedal a few times to center the shoes. Adjust the shoes until a firm drag is felt. Back off the star wheel seven notches to provide running clearance. Brakes that are adjusted too close will drag, heat up, and lock.

Repeat the above operation at the remaining wheels. Replace the cover plates; they keep out the road dirt.

To adjust the parking brake, first check the clearance between the idler level and the mounting bracket to see that the front cable is properly adjusted. This clearance from the rear of the idler lever to the mounting bracket should be 1/4 of an inch. To correct, disconnect the equalizer from the idler arm. The adjustment is made with the check nuts around the clevis.

To adjust the rear cable, pull the hand brake lever out seven clicks (not notches as they are not the same). Loosen the forward check nut on the equalizer and tighten the rear one until there is a slight drag on both rear wheels when rotating the drums. Tighten the check nuts securely. Set the parking brake lever back to two clicks from the full-release position. The brake shoes should be free. The brake lever should operate freely. If it is sticky, lubricate the pulleys and the parking-brake shaft. See that the equalizer return spring is not missing.

SERVICING THE MASTER CYLINDER

By removing the end bolt, disconnect the hydraulic line from the end of the cylinder. Cap the line with masking tape to prevent entry of foreign material. Remove the four retaining nuts and lock washers that hold the master cylinder to the firewall, and remove the master cylinder from the vehicle. Wash the outside of the body and empty out the old fluid. Don't reuse the old fluid. Disassemble the master cylinder completely, clean all the parts in alcohol, and blow dry. Do not use gasoline or solvent. Inspect the cylinder surface for scoring, pitting, and heavy corrosion. If any of these exist, the master cylinder is not rebuildable and will have to be replaced. Very light scoring can be removed with a fine polishing paper such as crocus cloth. Wash the bore with alcohol and blow dry. See that the openings for the inlet port and compensating port are open and the cylinder side is smooth.

Check the piston-to-cylinder clearance by placing the piston in the bore. Using a clean feeler gauge, measure the clearance. If the clearance exceeds .005 of an inch, the master cylinder is not rebuildable. If the clearance is fine, obtain a master cylinder rebuild kit. Do not reuse the old parts.

Wash hands with soap and water before you handle any of the new rubber parts. Install the piston stop and snap ring in the front of the master cylinder. Coat the cylinder bore with clean brake fluid. Dip the piston and primary cup in brake fluid. If you reassemble the cylinder dry, the parts might stick and score. Push the piston down into the bore so that the bleeder holes face upward. Slide the primary cup in with the flat side against the piston. Install the piston spring and the valve assembly. Dip the valve seat washer in brake fluid and fit it on the button of the end plug.

Assemble a new gasket over the end plug and screw the plug into the master cylinder. Tighten securely. Look through the reservoir and check that the compensating port is not covered by the primary cup. If it is covered, first check that you have the proper rebuild kit. Check the thickness of the new stop ring against the old one. If the old one is thinner, use it. If not, flat file the new ring or the piston end that sits against the ring to shorten this distance.

Place the master cylinder in the vise and fill the reservoir with brake fluid. Replace the filler plug and gasket. See that the filler plug vent hole is clear. Use a screwdriver and force the piston forth and back a few times until the air is forced out. Catch the brake fluid in a suitable container. Top up the reservoir. Install the rubber pushrod boot. Place the master cylinder in position on the fire wall with the pushrod in place in the piston. Tighten the mounting nuts.

Replace the end bolt in the end plug. There are two copper gaskets, one on each side of the connector, to prevent leaks. Leave the bolt a bit loose and wrap a cloth around the end of the master cylinder. Pump the brake pedal gently once to remove any remaining air. Tighten the end bolt before releasing the pedal to prevent drawing air into the system. It seems like you will need help on this one, but you could hold the brake pedal in down position with a stick or brace. Check the brake pedal free play and adjust as necessary.

You should have solid pedal. If this isn't the case, you will have to bleed the brakes. Low pedal can be increased by brake shoe adjustment, but a spongy pedal means air. Vehicles that use power brakes have a self-contained vacuum and hydraulic cylinder to reduce the amount of pedal travel and foot pressure needed for brake application. The unit utilizes engine vacuum and atmospheric pressure that acts on a diaphragm to apply pressure to the master cylinder plunger. In the event of engine failure, the brakes will still function, although more pedal pressure is required.

To check power brake operation, shut off the engine and apply the brakes several times to use up the vacuum in the reserve tank. Then depress the brake pedal and start the engine. If the pedal moves downward and less pressure is required to hold the pedal in applied position, the power brake system is working. If it does not, then the vacuum part of the power brake system is faulty. If the brake pedal continues to the floor, the hydraulic system is leaking.

Check the brake fluid level in the reservoir. It should be within 1/2 inch of the top of the filler cap opening. Inspect the lines and connections. Check the wheel cylinders. Inspect the hydraulic lines and connections attached to the power brake hydraulic cylinder output port for leaks. If the problem is in the hydraulic cylinder, the entire assembly must be removed for service. See that you can get parts or have the entire assembly rebuilt at a brake shop.

To check the vacuum booster, use a vacuum gauge at the booster and at the engine. The amount of vacuum should be the same. If not, check for cracked hoses or leaking connections at the engine intake manifold. Check the vacuum lines and connections at the vacuum inlet tube assembly, and check vacuum attachment at vacuum reservoir. When the engine is shut off, the vacuum should remain in the lines. Loss of vacuum indicates a leak. You might be able to obtain NOS or a rebuilt unit if the problem is in the vacuum booster.

The air cleaner should be cleaned at least twice a year. To clean, remove the air cleaner and wash thoroughly in solvent and allow to dry before reinstalling. Replace the filter if necessary.

A spongy brake pedal is caused by air in the system, and it will need bleeding. Do not run the engine while bleeding. The vacuum reserve should be exhausted by applying the brake several times with the engine off before starting the bleeding procedure. The procedure is the same as for the standard brake system. Start at the longest line first. Little brake pedal pressure might indicate the need for brake shoe adjustment or new shoes and/or drums. Same thing as for the standard brake system.

SERVICING THE WHEEL CYLINDERS

Once you have located the leaking wheel cylinder or when you are replacing brake linings, it is a good idea to service the wheel cylinders. Kits are available that contain new rubber cups and boots. New wheel cylinders will have to be used if the old ones are not repairable. As with all rubber parts, keep grease and solvent away. Use alcohol for cleaning and lubricate the parts with brake fluid for easy assembly.

Raise the vehicle on jack stands and remove the wheels. Pull the drums off. It might help to back off the brake adjustment. Remove the brake shoes and keep the parts in order. This makes it easier on reassembly doesn't it? It's possible to service these wheel cylinders on the backing plates. Keep them in place if you prefer. If you have to remove the rear ones, see that you have a flare nut wrench to remove the brake line or you will probably twist it off.

Remove the cylinder boots, pistons, rubber cups, and the spring. Wash all parts with alcohol. Inspect the cylinder bore and, if it is pitted or scored, it will need servicing. Before you do this, check the fit of the piston in the cylinder bore by using a feeler gauge. The clearance should be maximum .004 of an inch. More than this and the cylinders have to be replaced. Get new ones complete with cups and pistons. If this is NOS check to see that the cups and pistons aren't seized. Use brake fluid on reassembly.

If the wheel cylinder will take reconditioning, use crocus cloth or a wheel cylinder hone to clean up the cylinder bore. Light pitting in the middle of the cylinder is acceptable as long as the cups don't come into this area. Remove the bleeder screw and leave the bleeder valve loose in the cylinder.

Wash the cylinder with alcohol and blow dry. Dip pistons and rubber cups in brake fluid and coat the inside of the cylinder. Place the cups in the cylinder edges (facing inward and with the spring in between). Install pistons with the flat side against the cups. Snap the boots in place and replace the two connecting links.

Arrange the brake shoes with the primary facing frontward and place the adjusting screw in place so that the star wheel is in line with the adjustment slot in the backing plate. Connect the shoes with the adjusting screw spring. Pull the shoes apart at the top and attach them to the wheel cylinder connecting links. Replace the brake shoe hold-down springs and the brake shoe retracting springs. Tighten the bleeder valve. Replace the brake drum and the wheel. Adjust the brakes.

Check the master cylinder reservoir for brake fluid and bleed the necessary wheel cylinders. Try the brake pedal; it must be solid. If it is solid but goes past half travel, adjust the brake shoes at the other wheels. If this does not improve brake pedal travel, then relining or drum replacement might be needed. A spongy pedal indicates all the air is not out of the system. Bleed as required and remember to replace the bleeder screws. If the brake pedal goes to the floor, you have a leak in the system. Recheck all connections and wheel cylinders.

One last safety tip. There is a health hazard with asbestos and brake linings are made of this material. The cleaning of brake assemblies using air can produce airborne particles of asbestos dust. Breathing this dust can cause cancer. Use a vacuum cleaner if you decide to clean brake assemblies.

Chapter 8

Clutches and Transmissions

The clutch disconnects the engine from the rest of the power train. It is operated by a clutch pedal located in the driver's compartment. When the clutch pedal is up, power is being transmitted to the transmission. When you step on the clutch this connection is broken. The clutch assembly consists of four main parts: the flywheel, the pressure plate, the disc, and the control pedal and linkage.

A single-plate, dry-disc type of clutch is used on all standard transmission models. The turning effort of the crankshaft is transferred to the flywheel. When the flywheel has reached a certain speed, its turning effort is transferred to the transmission shaft. This is done through the disc that is pressed against the flywheel by the pressure plate and it begins to turn as one unit. The disc is splined to the transmission input shaft and the rotary motion of the crankshaft is now transferred to the gears.

It takes a little practice to operate the clutch pedal so that the clutch will be engaged smoothly and the vehicle will start out the same way, but you can do it. When the clutch starts to slip or it becomes difficult to shift gears without grinding, it probably means the clutch needs replacement. A squealing sound when the pedal is depressed indicates a bad bearing. Chatter can be caused by a rear seal oil leak. All the above mean that the clutch is trying to tell you it's time for service.

Without taking out the clutch, the only service you can provide is the adjustment of the clutch linkage. This is known as clutch pedal free travel and compensates for the wear on the linings of the disc. It sets the proper clearance between the clutch-release bearing and the diaphragm spring fingers. As the linings wear, the pressure plate moves in toward the flywheel and the clutch pedal free travel is diminished. The clutch pedal should travel up to 1 inch before the clutch-release bearing engages the diaphragm spring fingers. Check this movement by hand and not by foot.

To adjust, raise the front of the car and set it on jack stands. Don't get under with only the support of a jack. A hydraulic lift can come down on you faster than a barber's chair. Locate the clutch assembly; it's right behind the motor. Now look for the clutch release fork. It is sticking out of the clutch housing. There is a threaded rod with two nuts on it that fits into the clutch release fork. That is what you are looking for. Loosen the lock nut and turn the adjusting nut until the clutch pedal free travel is about 1 inch. This keeps the clutch release bearing about 1/8 of an inch away from the diaphragm spring fingers. Tighten the lock nut.

If you find that the free travel is difficult to feel, you can decrease the pull-back spring tension. It should take 6 to 9 pounds to move the pedal off the stop. This adjustment is under the dash. Remove the rubber bumper from the clutch pedal return stop bracket on the right side of the dash panel. Let the clutch pedal come up against the stop bracket. Insert a pry bar between the underside of the dash panel and the top side of the clutch-pedal, return-spring lever. Loosen the two clutch-pedal-to-clutch-pedal, return-spring-lever bolts.

Decrease or increase the spring tension as required and retighten the bolts. Depress the clutch pedal and install the rubber bumper. Recheck the clutch pedal free play travel. If you find that the height of the clutch pedal does not match that of the brake pedal, there is an adjustment by means of a slotted hole in the clutch lever bumper bracket. You might have to recheck the pull-back spring adjustment and the pedal free-play adjustment.

REPLACING THE CLUTCH

Before you remove the clutch, make a few simple checks to determine if the clutch needs replacing. Make sure the clutch pedal free travel is alright. Check the clutch pedal bushing for sticking on the shaft or loose mountings. Lubricate the pedal linkage at the bell crank. Tighten all front and rear engine mountings (replace if oil-soaked). Check the transmission mounting bolts. See that the rear spring shackles are tight. Check the U-bolts. Check the clutch bell crank between the engine and frame for sticking and looseness.

Still slipping, grabbing, or chattering? Then it has to be the clutch. The transmission has to come out first. Raise the vehicle on stands (two are good, four are better). Drain the lubricant from the transmission. This is not to make it lighter but to keep you cleaner when you remove it. Disconnect the speedometer cable, and the shift control rods from the levers at the transmission. Undo the rear universal joint U-bolts and pry the universal joint off the pinion drive flange. Pull the drive shaft out of the transmission. There might be slight oil loss. Remove the two top-transmission-to-clutch-housing-attaching bolts. If you are working by yourself and there is no one to help you remove the transmission, make up two transmission guide pins. Use regular bolts or a threaded rod that will fit into the threaded holes in the clutch housing. Saw off the bolt heads and cut a screwdriver slot in the ends. Lengths of about 6 inches are fine. These same guide pins will also make it easier to install the transmission. They prevent the weight of the transmission from hanging on the clutch disc and springing it.

Remove the two lower attaching bolts and slide the transmission straight back on the guide pins until the clutch gear is free of the splines in the clutch disc. The transmission is not too heavy but it is awkward working. Place it out of your way. Remove the clutch release bearing from the fork. Remove the clutch release bearing from the fork. Remove the clutch fork tension spring from the fork and disconnect the clutch fork pushrod. Remove the clutch fork by forcing it forward and toward the center of the clutch. Remove the clutch housing underpan.

Loosen the clutch attaching bolts a bit at a time until the diaphragm spring tension is released. Then remove the bolts and remove the clutch disc and pressure plate. This is fairly heavy so don't let it fall on you. Examine the flywheel clutch surface. If it is scored it should be reground. Very light heat cracks are permissible but the glaze should be broken with emery cloth. Sand across the surface of the clutch disc contact area. Check for an oil leak at the rear seal and repair. Oil on a clutch surface is the first step toward slippage. Examine the pilot bushing for excessive wear. This bushing supports the front end of the transmission input shaft and can cause clutch chatter.

To remove the bushing, use an expandable puller or thread the bushing to fit a bolt so that you can pull the bushing out. Drive the new one in by using the old bushing as a driver. Lubricate the inside of the bushing with high-temperature grease. Inspect the pressure plate for excessive burning, heat checking, warpage, and scoring. Check the diaphragm spring for cracks, overheating, and looseness. Check the ends of the release fingers for wear. Inspect the clutch disc for worn, loose, or oil-soaked linings, broken springs, loose rivets, and worn hub splines.

Which should you replace? Both. There is a lot of work to replace the clutch once so why do it twice? You might be able to get by with a clutch disc but its better to replace disc and plate. These are sold as exchange units and it comes out less expensive in the long run. Regarding the clutch release bearing don't even consider using the old one. You are asking for trouble. Keep greasy fingerprints off the new friction-contact surfaces. Wash the flywheel surface with a nonpetroleum base cleaner. See that your hands are clean.

Before you start, it would be nice if you had the use of a clutch-aligning arbor or a spare-input shaft. This would help align the clutch disc with the pilot bushing and make transmission installation very uneventful. However, with a bit of careful "eyeballing" you should be fine.

Turn the flywheel until the "X" mark is at the bottom. If there is a corresponding mark on the pressure plate cover, they should be aligned. Place the clutch disc against the flywheel with the correct side facing out. Place the pressure plate in position and start two bolts. Heavy work isn't it? Take a rest.

Start the remaining bolts and move the clutch disc so that it is in an even amount around the pressure plate. Tighten each bolt a turn at a time to prevent distorting the cover as the spring pressure is taken up.

Pack the clutch fork ball seat with a small amount of high-temperature grease. Replace the clutch fork on the clutch fork ball in the clutch housing. Lubricate the recess on the inside of the clutch-release bearing collar and coat the fork groove with some of the same. Install a clutch release bearing to the clutch fork and hook up the linkage.

Clean the splines on the transmission input shaft but do not oil them. The clutch disc must slide freely on these splines. If there is any sign of leakage through the input shaft bearing, correct it. Apply a thin coat of lubricant to that portion of the input shaft bearing retainer that supports the clutch release bearing collar.

Lift the transmission in place and support it on the guide pins. Push the transmission forward so that the input shaft and the bearing retainer pass through the clutch release bearing collar and the input shaft hits the clutch disc.

Place the transmission in gear and turn the output shaft so that the splines on the input shaft align with the clutch disc hub. Keep pushing forward. The front of the input shaft must enter the pilot bushing. Jiggle the back of the transmission to permit easier entry. This is where you find out if the clutch disc was aligned properly with the bushing. Not entering? Don't panic. Get a helper to turn the engine over with the starter while you keep steady pressure on the transmission pushing forward. Push in until the transmission touches the clutch housing. Sometimes stepping on the clutch pedal will help.

Install the two lower transmission mounting bolts and lock washers, and tighten securely. Remove the guide pins, put the upper bolts in place, and tighten securely. Place the drive shaft into the rear of the transmission and align the universal joint with the flange. Install the U-bolts and tighten the nuts securely. Connect the shift linkage.

Remove the speedometer driven gear and add 1/2 pint of transmission lubricant to the housing. Use a plastic squeeze bottle or a small hose and funnel. Reinstall the driven gear and connect the speedometer cable. Fill the transmission with lubricant. Lubricate the bell crank with chassis grease. Check the clutch pedal free travel and adjust. Start the engine and engage and disengage the clutch a number of times to seat the clutch disc. Recheck the clutch pedal free travel. Take the vehicle off the stands and you are ready to roll.

STANDARD TRANSMISSION SERVICE

Before you remove the transmission because of a noise problem, check the level of the lubricant. If the transmission jumps out of gear, check the shift linkage and the shift cover. Loss of lubricant could be past the rear seal or the shift-cover gasket. This work can be done with the transmission in the car. Hard shifting and gear clashing can be caused by excessive clutch pedal free travel.

With the vehicle on stands, drain the transmission lubricant and remove the shift cover. Turn the gears over slowly and inspect the gear teeth for chipping and excessive wear. Rock the gears on the shaft to determine clearance. Check end play of the cluster gear, reverse idler gear, and input and output shafts. These clearances should not be any greater than a couple of thousandths. Check up and down movement of cluster gear, reverse idler, and the input and output shafts. Trace out the power flow of the gear position that is causing the noise. If it is noisy in all gears, check the input and output bearings and the cluster gear.

Examine the shifting forks on the shift cover for wear and replace or build up with brass. Check the fit of the shift levers in the shift cover. You may find it necessary to replace the cover if this transmission has the habit of locking in two gears at one time. Check the interlock. If your decision is to remove the transmission, proceed as in the clutch section.

To replace the main shaft oil seal, the drive shaft will have to be removed. Examine the front flange for grooving because a new oil seal will not seal for long if the oil gets past the groove. Replace the flange if this is the problem. There is also a bushing in the transmission that supports the front flange. If it is worn excess oil gets past. Replace this bushing. Rent a puller to make this job easier. To remove the seal, use a puller or pry the seal out.

Clean the extension housing where the new seal is to fit. Coat the seal outside diameter with a thin coat of nonhardening sealer. Drive the seal in with the seal lip facing into the transmission. Use a pipe as a driver but don't damage the seal. Drive it in solid. Replace the drive shaft. Replace the shift cover gasket if a leak is indicated here.

To adjust the gearshift linkage on the 1955 models, move the shifter rods until the transmission is in neutral. The first and reverse shifter tube lever must be in the center of the slot. Adjust by loosening the swivel nut and moving the first and reverse lever. Tighten the nut. Align the second and third shifter lever with the first and reverse lever. Adjust by loosening the swivel nut and moving the second and third lever into position. Tighten the swivel nut. The shifter tube levers must be in line or key will not enter keyway in levers.

To adjust the 1956-57 linkge, make up the keyway tool as shown. Loosen the swivel nuts and insert the tool in the keyway in the adjusting ring at the bottom of the mast jacket through the first and reverse lever and the second and high lever. Move both control rods until transmission is in neutral. Tighten swivel nuts and remove the keyway tool. Start the vehicle and shift through the gears and check jumping out of gear, hard shifting or gear clash during shifting. If these still continue the transmission will have to be removed for service. Remember to have the transmission lubricant at the proper level, when doing the above tests.

OVERDRIVE SERVICE

The overdrive unit is attached to the rear of a standard transmission, and it provides for a lower engine speed while maintaining the road speed. Read the operator's manual so that you can operate this unit properly. There are two separate controls for the overdrive unit; the mechanical and the electrical. Before you remove the overdrive unit, do some checking to see why the overdrive unit will not engage.

The electrical circuit consists of three different circuits within the overdrive circuit: the control circuit, the solenoid circuit, and the ground-out circuit. First check for a blown relay fuse. Before you replace it check for the reason it blew. Check for broken wires or loose connections anywhere along the circuit. Raise the vehicle on stands and identify the electrical overdrive connections at the back of the transmission. They must be on tight and not corroded or broken.

The mechanical control moves the overdrive unit into the engaged position. If a buzzing noise is the result when the dash control unit is pushed in, then the control cable needs adjusting. Raise the car on stands and identify the control cable at the transmission. It is on the driver's side and is held by a bracket with the free end attached to the control lever. Loosen the binding post at the control lever so that the wire is loose. Pull the dash control knob out 1/4 of an inch and move the control lever all the way back. Tighten the binding post.

If you still have problems, have the electrical circuit checked out at a transmission shop. If tests definitely indicate the overdrive unit needs repairs, you could service this unit while it is on the car. On convertible models, the transmission and overdrive unit must be removed as one. On others, the over drive unit can be worked on by removing the overdrive housing and leaving the transmission in place. If the transmission main shaft, overdrive adapter or transmission rear bearing need replacement, the entire transmission will have to be removed. Check with the transmission shop.

AUTOMATIC TRANSMISSION SERVICE

Two different transmissions are used in these model years. The Powerglide and the Turboglide. A shift control lever is mounted at the top of the steering column so that the driver can select the necessary operating ranges. The Turboglide transmission uses a Hill Retarder (HR) range so that the engine can be used for braking. On the Powerglide the Low (L) range can be used for the same purpose. On Turboglide, the low range is built into the Drive (D) and a separate range is not used. Both transmissions can be started in either Park (P) or Neutral (N) range.

To push start, place the control lever in the Neutral (N) position until the car reaches 25 to 30 miles per hour. Then place the control lever in Low (L) or Hill Retarder (HR). After the engine starts, move the control lever to Neutral (N) for engine warm-up. Towing is not a safe method for starting as the vehicle might run into the back of the towing vehicle.

If it should prove necessary to tow a vehicle home, it would be best to do so with the rear wheels off the ground. If this is not possible, remove the drive shaft before towing. Short distances can be covered with the transmission in Neutral (N) and towing speed under 30 miles per hour.

The automatic transmission will provide miles of trouble-free driving under normal conditions as long as the oil level is kept at a safe operating range. The oil level can be checked once a month as part of your monthly service. The level can only change if the transmission has an external leak or if the oil cooler is leaking and coolant is entering. The dipstick level will indicate what is happening.

Have the vehicle in level position, motor at operating temperature with the control lever in Neutral (N), idle, and the parking brake on. The oil must be hot for an accurate reading. The oil dipstick tube is located in the engine compartment on the right-hand side by the starter. Wipe the area around the tube before removing the dipstick. On Turboglide, place the control lever in Drive (D). Check the level it should be at or just below the Full mark.

Examine the color; it should be reddish. If it is black or brown and has a burned smell, the transmission will soon need service. Wipe the dipstick between your thumb and forefinger and check for particles. If there are any, it confirms that service is needed. If the fluid is milky and above the Full level, coolant has entered through the oil cooler. Air bubbles and/or over Full level indicate foaming. This will cause faulty transmission operation, noise, and overheating. The excess oil will be forced out the dipstick tube. Drain the oil out and fix the problem.

If you find the oil level is low and the transmission is at the operating temperature, check for external oil leaks. The rear seal could be the problem. This can be fixed with the transmission in place, similar to service for a standard transmission. Check all plugs and bolts; they must be tight. If the transmission is dirty on the outside, wash it clean and then investigate for leaks. If oil is leaking out of the converter housing, it could also indicate an engine oil leak.

Converter leaks usually mean that the transmission will have to be removed for service. Fix these leaks. The lack of oil will burn up the transmission and that repair is costly. To change the oil in a transmission does not change the oil in the converter. If the oil is contaminated, but the gearing is fine, you might consider draining the entire system. This includes the oil cooler. You want to get all the old oil out. If the old oil is fine and just needs changing, the new oil will restore the quality of the old and extend the useful mileage of the transmission.

Bring the transmission up to operating temperature and raise the front of the vehicle on stands. Drain the oil by removing the transmission drain plug (Powerglide) or the oil pan drain plug (Turboglide). Don't burn yourself. This oil is hot. After complete draining, replace the plug and lower the vehicle. Refill the transmission through the oil dipstick tube, pour slow and use a transmission funnel. The correct oil is Type "A" Automatic Transmission Fluid (Powerglide 4 quarts and Turboglide 3 quarts).

Start the engine and bring the oil up to operating temperature. With the selector lever in Neutral (N) or Drive (D), check the oil level. Add just enough to bring the level up to the Full mark. Do not overfill.

That is about all you can do for service. Leave the band and linkage adjustments to the experts. Don't try to repair something you know little about or it will cost you. If you find that the transmission needs repairs, you might take the old one out, have it rebuilt, and then re-install it. This is not a difficult job, but it is time-consuming and you should be able to save some money by doing it yourself. Do not attempt the job unless you have a wheel type of floor jack.

Raise the vehicle front and rear and set it on stands. Now, get under and have a look at the setup. You should be able to identify most of the parts and see how they are connected to the transmission. There are a number of pieces you will have to take off because they are in the way. Remember parts that break, twist off, or get lost will cost you time on assembly. Work methodically. You might think there is no time but you will have to find time if the job has to be redone. Work safely. This is no different than any service job. If for some reason you don't feel good about something, leave it alone.

Turn back the floor mat in the passenger compartment and remove the toe-pan plate. Through this opening, you will see three bolts; they come out last. Spray penetrating oil on the exhaust pipe flange on the six-cylinder model. Remove the spark plugs (on all models) so it will be easier to turn the engine over.

Disconnect the battery ground strap and the solenoid wires on the eight-cylinder models. Drain the transmission and the converter on the six-cylinder models. Less mess this way! Disconnect the oil cooler lines at the transmission and from the retaining clip on the engine. Cap the ends of the lines with plugs or masking tape. Move the lines away from the transmission and tie them to the frame rail.

Disconnect the speedometer cable and the control rods. Mark the rods with masking tape as an aid to assembly. Disconnect the vacuum hose on the Turboglide. Remove the dipstick tube. Split the rear universal joint and remove the drive shaft. Going pretty good? Take a break.

Remove the starter on eight-cylinder models. Disconnect the exhaust pipe from exhaust cross-over pipe on eight-cylinder models. On six-cylinder models, disconnect exhaust pipe from exhaust manifold. Disconnect the muffler from the support bracket. Move the muffler and exhaust pipe to the left and wire to the left frame side member. Remove the converter underpan (Turboglide). Remove the flywheel inspection hole cover (Powerglide). Remove the three flywheel-to-converter-attaching bolts through the opening in the flywheel housing that is adjacent to that of the starter on the eight-cylinder model (Powerglide).

On the six-cylinder models this opening is on the left side of the engine. On the Turboglide remove six bolts and nuts. It is necessary to turn the engine over so that all fasteners will be exposed. Do not pry on the ring gear or it may warp.

Slide the jack under the transmission and raise it into the transmission just so that the rear engine mounts are a bit relieved. Support the engine by placing blocking under the oil pan and building a trestle. Do not pile the blocks on top of one another. Do this right and prevent personal injury. Lower the transmission jack so the engine sits on the trestle. Raise the jack up to the transmission and secure the saddle to the transmission using a piece of chain. The transmission is fairly heavy so you need a solid mounting not only for removal but also for safe assembly.

You might consider taking a piece of 3/4-inch-thick plywood

and that is a foot square and securing it to the jack saddle. Use some short blocks to support the transmission to the plywood and nail them in place. Use a chain to hold the transmission in place on the plywood. Drill some holes and bolt the chain down. You might consider renting a transmission jack. Great. Remove the rear engine mounts on both sides. Remove the bolts that hold the transmission case to the engine block including the three reached through the toe pan. Check to see that all is loose. Pull the jack back enough so that you can lower the transmission. Wire the converter to the transmission housing so it doesn't slip off. Remove the transmission from under the car. Get some help to load the transmission for the ride to the rebuilding shop. Rebuild the converter also.

Installation of the rebuilt unit is pretty well a reversal of its removal. Align the "X" mark on the converter cover with the "X" mark on the flywheel to maintain balance. Grease the crankshaft bore to fit the converter cover a bit easier. Use care so that the converter does not slide forward and disengage the front pump drive gear.

Align the transmission with the engine block and the flywheel with the converter. Slide them together and engage the bolts in the engine block. Start all the bolts and bring them up tight. Install the converter fasteners while rotating the flywheel as necessary. Tighten bolts 25 to 30 foot pounds. Tighten nuts 15 to 20 foot pounds. Install the rear engine mounts and remove the blocking. Remove the transmission jack. Replace the linkages and lines. Connect universal joint. Connect the exhaust system and install the starter.

Install the spark plugs, the solenoid wires, and the battery-ground strap. Fill with 3 quarts of transmission oil and start the engine. Complete the filling with 8 quarts more for the Powerglide and 6 more for the Turboglide.

Move the control lever through the gears and check the level, bring to operating temperature, and add if necessary. Check for leaks while on stands. Lower the vehicle to the floor and road test. Re-check oil level and leakage. Everything should be fine.

Chapter 9

Driveshafts and Rear Axles

The drive shaft delivers the power from the transmission to the rear axle. It must flex and stretch as the rear wheels bounce over the rough roads. Universal joints at each end of the shaft allow it to change angles. A splined joint at the front slides in and out to allow for changes in length. The floor pan has a hump in the middle to provide clearance for the up and down movement.

The Chevrolet passenger vehicles use an open drive shaft known as the Hotchkiss drive system. The two exposed universal joints are of the sealed roller bearing type and are lubricated at assembly. The drive shaft itself provides a slight cushioning twist that helps to preserve the gear teeth in the transmission and the rear axle. The only service for the drive shaft is the replacement of universal joints or if the drive shaft is bent the replacement of the drive shaft. Vibrations or noise at low driving speeds are the first signs of a drive-shaft problem.

Raise the car at the rear and place stands under the rear axle housing. Slide under the vehicle and grasp the drive shaft near the universal joint. Twist the shaft in both directions. Any movement indicates the universal joint should be replaced. If the universal joints are good, check the drive shaft. Start the car and place the transmission in gear. Watch the action of the drive shaft; it should turn straight with no up and down movement. If the drive shaft is sprung or badly dented, replace it.

REPLACING UNIVERSAL JOINTS

To replace the universal joint(s), it is necessary to remove the drive shaft. If the universal joints do not have grease fittings on them, they will have to be disassembled at 25,000 miles and lubricated. The procedure is the same for both. Mark the position of the rear universal joint, in relation to the pinion flange, so that the drive shaft can be reinstalled in its original position.

Split the rear universal joint by removing the "U" clamps. If this bearing is good or only needs lubricating, tape the cups to the cross. Lower the shaft to clear the flange and pull the slip joint yoke out of the transmission. Remove the assembly from the vehicle.

Clean the road dirt out of the yokes and spray some penetrating oil around the cups. Remove the lock rings. If you have a fairly large bench vise, you can use it to press the cups out or you can use a soft drift and a hammer. Place a large socket or a short length of 1 1/4-inch pipe to support the yoke. Using the soft drift and hammer, drive the opposite cup out of the yoke. Support the other end and drive the cup out by using the soft drift against the cross. If you are working on the front universal, remove the cups from the slip yoke in a similar fashion.

Discard the worn universal joint and replace it with a new one. Never replace new cups on an old cross (or vice versa). If you are lubricating, wash all the parts and blow dry. If the rollers and the cups are free of corrosion and grooving and if the cross is all right, the parts can be lubricated and reused. Use high-melting-point wheel bearing grease.

Check the drive lugs on the drive shaft and the slip yoke. Check the slip yoke for scoring or grooving and replace or repair. Repair can be done with brass and a lathe used to true up the repair. If you believe the slip joint seal in the transmission needs replacing, this is the time to do it. Remove the old seal using a puller or pry bar. Clean the seal area of the transmission. Coat the outside of the seal with sealer and drive the seal into the transmission with the seal lip facing inward.

To assemble the universal joint, start one of the cups in the yoke lug with the open end up and the needles in place. Slide in the cross. If the cross has a grease fitting, it must face inward or you will not be able to fit the grease gun on. Start the other cup. Make sure it slips over the cross.

Drive or press the cups inward far enough to install the lock rings. Tap the cross with a soft-faced hammer to seat the cups against the lock rings. If one ring goes in and the other won't, it might be that a needle has fallen out. The universal joint will have to be taken apart and inspected. Replace the other universal joint in a like manner, and if it has a grease fitting keep it in line with the first one.

See that there is full range of motion in the universal joint. To re-install the drive shaft, see that the slip joint is clean. If a new seal was installed, lightly grease the slip joint. Slide it into the transmission so that the reference marks will align at the rear. If they don't align, pull it out and re-install. Connect the universal joint by installing the "U" bolts. Be sure the cups are properly seated in the pinion flange. Road test. If very slight vibration exists, try repositioning the slip joint or the pinion flange 180 degrees. Minor unbalance can be corrected by attaching a screw type of hose clamp to the rear of the drive shaft and experimenting with its location until the vibration is reduced.

REAR AXLE SERVICE

The rear axle does a number of jobs so that you can get final drive. It changes the engine torque to a 90-degree drive, it reduces the drive-shaft rpm, and it makes it possible to turn corners. With all these drives, there has to be some noise at certain speeds. It is only when the noise becomes unbearable at all speeds that we finally become alert to the problem. Sometimes axle noise is confused with tire noise, transmission noise, drive-shaft vibration, or universal joint noise. Eliminate these and you should only have the rear axle left. To eliminate tires, inflate both front and rear to 50 psi. The noise should now be altered. Reduce the pressure and continue with the other tests. Transmission noise should include tests at different speeds and gears. Check the level of the lubricant. Note the gear in which the noise is most evident. Raise the car and check for drive-shaft vibration and universal joint wear. Sometimes engine sounds will telegraph down the hollow drive shaft. With the engine running and car standing still, bring the rpm up. If you can hear the noise, then it's not in the rear axle.

Rear axle noise will be gear noise or bearing noise. Gear noises will be variable in pitch and will be most obvious in certain speed ranges and gear conditions such as drive, cruise, float, and coast. Bearings will cause a noise that is constant in pitch and varies over a wide range of road speeds. It will be the most noticeable in drive under acceleration. The bearings that can make noise are the pinion, wheel, and side bearings.

Pinion bearings noise is low pitched and continuous. They are rotating at a higher speed than the wheel bearings or the differential side bearings. Wheel bearing noise can be picked up when driving by turning the steering wheel sharply to the left and the right. If the noise increases in the turns then it is caused by wheel bearings. Noise caused by front wheel bearings can be reduced or altered by stepping on the brake pedal while maintaining road speed. Side bearing noise is a low growl of a slower speed than a pinion bearing noise. Check the level of the lubricant in the rear axle housing. It should be level with the filler plug when the axle is up to operating temperature.

Before you remove the differential assembly for repair, get an opinion (preferably) from the mechanic at the service shop that will be doing the repair. If you remove this assembly from the axle housing and replace it, you will have gone a long way toward saving money. Special tools and pullers are required to rebuild the differential, and you are a lot wiser to have it done that way.

Raise the vehicle on rear stands and remove the wheels. Remove the axle shafts. Mark and split the rear universal joint. Remove the drive shaft. Tape the universal together. Remove the fasteners that hold the differential assembly to the axle housing. Place a container such as a plastic bowl under the housing. Separate the differential assembly and the axle housing so that the lubricant will drain out. This is a heavy piece so have a good hold of it. Remove it from the vehicle. Repair and replace.

Installation is the reversal of removal. Clean out the housing and use a new gasket. Use new copper gaskets and nuts and tighten securely. Connect universal joint. Service axle shafts if required. Replace the wheels. Fill with proper hypoid lubricant. Road test.

REPLACING PINION OIL SEAL

Lubricant leaks can cause a major problem if they are not looked after. This seal will let a lot of lubricant through, and if the flange is worn the problem is compounded. Raise the rear of the vehicle and place stands under the axle housing. Separate the rear universal joint and mark for reassembly. Tape the cups to the universal joint. Remove the drive shaft. If you have an inch-pound torque wrench, measure the amount of torque required to turn the flange nut. Try to move the flange up and down. If you can it indicates worn or loose pinion bearings. The differential assembly should be removed for inspection. The noise would be pretty severe.

If you don't have a torque wrench, count the number of threads exposed beyond the nut. This will aid in correct reassembly. Punch mark the position of the flange in relation to the position of the pinion.

Remove the pinion nut and the special washer. Spray some penetrating oil into the splines. To remove the nut, you should use a 3/4-inch drive set; it's on plenty tight. You will also have to hold the flange. You can try a large pipe wrench but secure it so that it can't slip. See if you can rent a flange holding tool because it is much safer. Use a hammer and carefully drive the flange off by alternately hitting the ends. Do not bend the flange because the universal joint will not fit in. Examine the flange for smooth oil seal surface, undamaged deflector, and unworn drive splines. Replace or repair if necessary. Order a new seal and a new self-locking nut.

Pry out the old seal. If the new seal is leather, soak it in light engine oil for 10 minutes. Lightly oil the splines on the pinion and the flange. Wipe off the outer diameter of the seal and coat with nonhardening sealer. Drive the seal into the carrier (with the lip facing inward). Use a short piece of pipe as a driver.

Install the drive flange (aligning the marks). Do not pound on the flange because you will damage the bearings or gears. Install the washer and the new nut, and use the nut to pull the flange on. Rotate the pinion shaft a few times to seat the bearings. Tighten the nut to remove the end play and continue tightening until the reference marks line up and the same amount of threads are exposed. Check with the torque wrench until the reading is the same as when you started. Do not over tighten.

Replace the drive shaft and universal joint. Lower the vehicle and check the lubricant level. You can add oil by using a plastic squeeze bottle. See that the filler plug is tight. Road test the vehicle.

REPLACING REAR WHEEL BEARINGS

This service is also the one to follow when replacing rear-axle oil seals. Raise the rear of the vehicle and place the stands under the rear axle housing. Turn the wheels slowly and listen for noise. See if a roughness can be felt. It might be necessary to remove both axles for bearing inspection. Remove the hub nuts and remove the wheels. Remove the brake drum and gasket. If the drum comes off only part way, loosen the brake shoes. Check for leaking brake fluid by pulling back the boots at the wheel cylinders. There should not be any more than a trace of wetness here. Any more and the wheel cylinders are probably leaking.

Remove the four nuts and washers that hold the bearing retainer to the backing plate. Attach a puller of the slide hammer type to the axle flange and pull the axle bearing free of the housing. If the bearing retainer and the brake strut interfere, raise the strut slightly with a screwdriver to obtain clearance. Install a bolt and nut to hold the backing plate in place.

Wash the axle and inspect the splines for wear. Check for cracks and evidence of twisting. Check for excessive runout. See that the flange bolts are not broken, stripped, or loose. Inspect the bearing for looseness, roughness, or evidence of leakage at seal. If you are going to replace the bearing, the seal, or both, it's better and much faster to use a press. Order the parts as required: seal, bearing, "O" ring, retaining ring, gasket, and flange bolts if needed. Now go to the press works.

Using a sharp cold chisel, notch the bearing retaining ring and slide it off the axle. Press the bearing off the shaft. If you are only replacing the oil seal, pry out the old one with a screwdriver and install a new one. Tap it in place. Make sure the bearing retaining is in place and install the bearing. Press it down until it seats against the shoulder. Install a new axle shaft bearing retaining ring on the shaft with the chamfered side against the bearing.

Replace the necessary flange bolts and press them into place. Keep the heads tight and use a punch to peen the shoulder of the bolt into the countersink around the bolt hole in the flange. This operation might be easier to do before you press the bearing on. Just a thought. Put them in the right way.

To install the axle shaft, remove the bolt holding the backing plate in place. Replace the bearing "O" ring seal. Install a new retainer gasket. Pass the axle into the housing and align the bearing and the splines. Drive the bearing in until it seats against the shoulder in the axle housing. Raise the parking brake strut if the bearing retainer needs clearance. Align the bearing retainer so that the oil drain section is over the drain hole in the backing plate.

Install nuts and lockwashers on bearing retainer bolts and tighten securely. Install the brake drum gasket, drum, and wheel. Adjust the brakes if necessary. Do the other axle. Lower the vehicle, check the oil level, and road test.

There is a gasket between the differential assembly and the axle housing that, if not properly tightened, will cause a lubricant leak. Once it starts to leak, further tightening will not help and the gasket will have to be replaced. Raise the rear of the car on stands. Use a wire brush and clean the gasket joint area. Remove the axle shafts as in the last service heading, but pull them out only far enough to clear the differential side gears. Split the rear universal joint and tape together. Remove the nuts and copper washers that attach the differential assembly to the axle housing. Use a plastic dish pan to catch the oil when the assembly is moved forward. Remove the assembly.

Clean the axle housing and the gasket surface. Check for a crack or a loose bolt that would let the lubricant by. Check the gasket surface on both parts. Place the new gasket over the carrier mounting bolts. Install the differential assembly to the axle housing. Use new copper gaskets and nuts to tighten evenly and securely. Connect the universal joint. Replace the axle shafts and wheels. Lower the vehicle. Fill with lubricant to the proper level. Check for leaks.

Very rarely would you have to replace the axle housing, but in the case of an accident that would cause distortion of the axle housing and the axle tubes it would prove necessary. You might also want to change the housing and the differential assembly as a unit if you were changing component parts from one chassis to another because of body style. If the springs are good, leave them on the axle housing. Raise the rear of vehicle and place stands under the frame side rails (*not* the axle housing). Have the tires off the floor so that the spring will hang in normal position. You want the vehicle high enough so that when you loosen the spring you can roll the entire assembly out from under the car using the wheels. Place the floor jack under the rear axle housing and raise it to relieve the tension of the rear shackles.

Remove the two self-locking nuts from the rear shackle pins and the outer shackle plate. Use a pry bar and rotate the spring eye around the hanger eye to provide clearance for the shackle removal. Drive the shackle off. Do the same for the other shackle. Disconnect the hydraulic brake line connection at the rear housing. Disconnect the hand-brake cable at the equalizer and remove the cables from the cable clamps on the frame. Split the rear universal joint and tape it together. If you lose one of the bearings, you will have to replace the entire universal joint because these bearings are not sold separate. Disconnect the shock absorbers from the anchor plates. Remove the front bushing nut and bolt. Guide the bolt out with the socket and extension or you will lose it in the frame side rail. Lower the jack and roll the rear axle assembly out from under the car.

Three different ratios are used on the rear axle (depending on the type of transmission). Standard ratio is 3.70:1, Overdrive is 4.11:1, and Powerglide is 3.55:1. In the 1957 model, the Powerglide and the Turboglide ratio 3.36:1 is and the Standard ratio is 3.55:1. The 1957 models have a drain plug in the bottom of the axle housing. You can get a pretty good idea of the ratio by marking the pinion and the tire, and then turning the pinion until the tire has moved exactly one full turn.

If the pinion moves about 3 1/3 turns, the ratio would be 3.36:1, about 3 1/2 3.55, 3 3/4 3.70, and 4+ 4.11. A large letter P cast into the carrier indicates a Positraction differential. A ratio around 3.55:1 is good for highway and street driving. If you do a lot of in-town driving, you might be interested in a 4.11:1 gear ratio.

Roll the exchange assembly under the car and connect the front of the springs. Leave the nut loose for now. Raise the rear of the spring so that the shackle can be installed, and then let it down to normal position. Install shackle outer plate. Install nuts and leave loose. Do the other side. Install the shock absorber and tighten securely. Connect the hydraulic brake line. Connect the hand brake cable. Reassemble the universal joint. Make sure the nuts are tightened evenly and securely. Bleed the brake system; you might be able to get by with just the right rear cylinder. Make sure there is a good, hard pedal on the first push. Lower the vehicle and bounce it several times. Tighten the shackle nuts 25 to 30 foot pounds and the front bushing nut 75 to 90 foot pounds.

There is one more service you can do, and that is the replacing of the axle when it is broken. Finding the problem is easy enough; the car won't move. Raise the vehicle with stands under the axle housing. Locate the broken axle. Remove the wheel and brake drum. Remove the nuts and washers from the bearing retaining bolts. Pull what is left of the axle out. Fish out the other pieces by using a magnet or a wire hook. If a piece is still left in the axle gear, pull it out by snaring it with a wire loop. Put all the parts together. If the break is clean and no parts are missing, you are probably fine. If some parts are missing, you might have to remove the differential assembly from the axle housing the flush everything clean. Even one small chip can ruin the bearings and gears. Don't take a chance on it.

Check the bearing and see if it can be reused. Get a new oil seal and "O" ring and retainer gasket. Get a new axle with bearing if necessary. You will need a new bearing retaining ring. The chamfered side goes toward the bearing. Slide the seal on and press the bearing in place (chamfered side to the flange). Press on the bearing retaining ring. Replace the bearing "O" ring seal.

Install the new retainer gasket. Slide the axle in until the bearing seats against the shoulder in the axle housing. Install the nuts and lock washers on the bearing retainer bolts and tighten securely. Install the gasket on the axle flange and the drum. Adjust the brake shoes if necessary. Install the wheel and tighten the wheel nuts. Lower the vehicle. Check the lubricant level. It should be at the bottom of the filler plug hole. Overfilling causes blown grease seals at the wheels. Lubricant capacity is 3 1/2 pints.

Cars that use Positraction differential transmit power to both wheels. Do not raise one wheel and engage the transmission because the vehicle will drive. To check the type of differential, raise the rear of the vehicle and turn the tire by hand. If the other side rotates in the *same* direction, the vehicle is equipped with a Positraction differential.

Removal for service is similar to the standard type. Have the repairs done at a service shop.

Do not use hypoid gear lubricant in this type of differential. You must use the specified oil that is available from the dealer.

Chapter 10

Cooling and Exhaust

The engine develops a great amount of heat. Only about a third of this heat/energy goes to drive the wheels. The exhaust wastes about a third and the engine absorbs about a third through the cooling system. This must be controlled so that the engine does not operate at too high a heat but at a temperature that will produce the most efficient operation. The Chevrolet car used a liquid cooling system.

Hollow passages called water jackets surround each cylinder. The heat energy is conducted through the cylinder walls and is transferred to the liquid in the water jackets. The water is circulated out of the water jackets and into the radiator—that—acts as a heat exchanger, and is then returned to the water jackets to pick up more heat. This circulation of water maintains a safe operating temperature. An efficient water temperature is maintained by controlling the water circulation by the use of a thermostat (a heat-controlled valve). When the engine is cold, the thermostat valve is closed—stopping the circulation. As the engine warms up, the valve opens and maintains the preset temperature. The units of the cooling system include the water pump, radiator, radiator hoses, fan, fan belt, thermostat, and radiator pressure cap. In all areas where the temperature drops below freezing, it is necessary to add antifreeze to the cooling system. Add sufficient antifreeze to provide full protection.

COOLING SYSTEM CHECKS

The amount and condition of the antifreeze should be checked periodically. The level of the antifreeze should be above the radiator when the radiator cap is removed. Use a mixture of 1 part antifreeze to 1 part water. Adding water only will dilute the strength of the antifreeze. There should not be any rust deposits along the neck of the radiator.

If there are rust deposits, it means that the rust inhibitors have deteriorated, the system should be flushed, and new antifreeze should be put in. The first check that of coolant loss is generally caused by a leak in the system. This could be in the water pump seal, broken hose, rust through at the radiator, heater core, or the frost plugs. The coolant could also be leaking into the combustion chamber and is being burned.

Remove the radiator cap and with the engine running see if there are any bubbles coming up in the coolant. If there are bubbles, have the radiator shop do a test. Oil in the coolant—especially on a car that has an automatic transmission—could indicate an oil cooler leak.

The first time you notice a cooling problem will be when the radiator coolant starts to boil. If possible, get off the busy road and pull over to the side. Turn off the engine and open the hood. Do not remove the radiator if steam is coming out of the overflow pipe. Do a visual inspection of the hoses and belts and the front of the radiator. Is it clear? Look under the car. Is coolant leaking out? Now, here is the toughest part of all. Let the engine cool off for 15 minutes.

If you see what is wrong and you can fix the problem, then go and get the necessary repairs. Lock your car before you leave. Remember you have to bring back some coolant. A burst hose might be fixed with some tape, but if you don't have coolant your still not going anywhere. You might be able to improvise for a fan belt but what are you going to use for coolant? If you don't think that you can fix the problem, send for a tow truck.

You now have the car at home and it has probably cooled down. Check for leaks, burst hoses, or a missing fan belt. Grab the fan and shake it up and down. If there is excessive play, the bearing is faulty and the entire pump should be replaced. If the pump is leaking, the seal is damaged and the pump should be replaced. A rebuilt pump will do the job and its less expensive than a new one. The old one is a trade-in.

A leaking radiator will have to be removed for repair and cleaning. A blocked radiator can cause an overheating problem. Have the radiator shop clean and check it over. You remove it, they repair it, and you install. A radiator that is not repairable will have to be replaced. You can flush some of the dirt and rust out of the radiator by backflushing. The radiator is probably empty so this is a good time to do it.

You will have to install a flushing T in the heater inlet hose, and it is acceptable to let it stay in line (unless you are a 100 percent purist). These kits are available at the auto parts store. Locate the heater inlet hose. It is the one from the engine block *not* the one from the water pump. Install the T in the hose. Backflush as per instructions on the package. Start the engine and remove the radiator cap. Increase the engine speed to about 2500 rpm. If the radiator overflows, it is plugged and will have to be removed and flow tested at a radiator shop. If the engine still overheats after backflushing and if everything else looks fine, have a look at the thermostat. Drain about a gallon of water out of the radiator. The thermostat is located at the top of the engine at the end of the upper radiator hose.

Remove the two housing bolts and remove the thermostat. Bolt the housing back on. If the gasket is damaged use some silicone to seal the tear. Start the engine and let it reach operating temperature. The results should be self-explanatory. Replace the thermostat and the housing gasket. Drain out the rest of the water and put in a 50-50 mix of antifreeze and water. All models have a capacity of 13 1/4 quarts. Add 1 quart for any model with a heater. See that the radiator cap is of the proper pressure. This pressure cap allows the engine to operate at higher engine temperature without boiling the fluid. Inspect the seal on the pressure valve. It should not be cracked. If it is cracked, replace the cap with one of rated for 7 pounds pressure.

REPLACING A WATER PUMP

If there is a circulation problem and overheating still results, you probably have a defective water pump. Centrifugal pumps are driven by a fan belt. Capacity for the six-cylinder engine is 55 gallons per minute at 4000 rpm and for the eight-cylinder engine capacity is 44.5 gallons per minute at 4000 rpm. Check the fan belt tension regularly. If the belt is cracked, brittle, oil soaked, badly glazed, or slippery, it should be replaced. This work is much better done at home rather than on the road. The tension is adjusted by the generator movement at the slotted bracket. With light thumb pressure applied midway between the water-pump pulley and the generator pulley, the belt should deflect 5/16 of an inch on six-cylinder models and 13/16 of an inch on eight-cylinder models. Perhaps the belt is slipping and the water pump cannot do its job. Battery not seem as peppy as it should be? Generator not charging because of that same problem? A loose fan belt? Are you sure its a water pump problem? Everything else checked?

Drain the radiator and remove the water inlet hose from the pump. Loosen the bolt at the generator slotted bracket and push the generator inward to the engine so that you can remove the fan belt. Remove the heater hose from the pump housing. Remove the bolts that attach the pump to the engine block. Remove the pump from the engine. Take the fan and pulley off the pump hub. Inspect the fan for bent or cracked blades. Inspect the upper and lower radiator hoses. They should be replaced if they are oil soaked, cracked, leaking, or feel spongy or hard when squeezed. The same goes for the heater hoses. Its better to head off any problems rather than have them on the road. Replace the hose clamps if they will not tighten up properly. Take the pump and hose(s) to the auto parts store and buy the necessary replacements.

Install the pump pulley and fan on the pump hub and tighten the bolts securely. Use a new gasket on the block and install the pump. Tighten the bolts evenly. Install hoses and a fan belt. Fill the cooling system with coolant mixture and top up with antifreeze (not water). The more water you put in the more you dilute the mixture. Check the radiator core-to-fan clearance. It should be 5/8 of an inch to 3/4 of an inch at the point of minimum clearance. Adjust by adding shims. Start the engine. If you continue to have a heating problem, the radiator will have to be removed for commercial cleaning.

Just one more check before you proceed. Use a timing light and check the ignition timing. Set if necessary TDC on six-cylinder models and 4 degrees BTDC on eight-cylinder models. Late ignition timing can cause engine heating.

If you find that the temperature indicator stays in the "hot" range, the entire unit will have to be replaced if it is of the capillary type. On the electrical type indicator, check the wire leading from the gauge to the engine unit to see if it is shorting to ground. If not replace the engine unit.

To remove the radiator for repair or replacement, first drain the coolant. Remove the oil cooler lines from the radiator and cap the ends with masking tape to prevent dirt entry. This is only for automatic transmission models. Remove the water outlet hose and the inlet hose. If they will not pull loose after the clamps are released, cut them and peel them back. Remove the radiator core to the radiator support bolts and lift the core straight up to take it out. Save the shims. Service the radiator. Depending on its condition, it might be wise to take the replacement route.

When installing the radiator, replace the shims to provide the necessary core-to-fan clearance. If applicable, install the hoses and connect the cooler lines. Check the radiator cap and refill with coolant mixture. Start the engine and check for leaks.

Coolant leaks are a problem that you should correct as soon as you find out where the source. The core plugs start to leak as the inside of them rust out. Change these as soon as you can. Examine the outsides for any sign of rust that indicates the plug is starting to rust through. The ones on the six-cylinder engine are easier to replace but they all include a lot of work. The parts cost very little. A leaking heater core is best replaced. It is not practical to repair it. A leaking head gasket(s) is strictly a case of replacement. A cracked cylinder head(s) is also a case of replacement on your part. Do what you can do best and leave the mechanics with the rest.

EXHAUST SYSTEM

The exhaust system used on all cars consists of an exhaust pipe, muffler, and tail pipe that discharges the exhaust gases. The V-8 engine models use a cross-over pipe to connect the two exhaust manifolds. Some V-8 models use a dual muffler exhaust system. There is no such thing as fixing the exhaust system. The only thing you can do is replace the rotted parts. Noise is the dead give away that the exhaust system needs looking at. Next to noise is the silent killer known as carbon monoxide. Don't take any chances; you are flirting with death.

Usually the further you get from the engine heat the sooner the parts corrode because the engine heat eliminates moisture. Road salt used during winter driving conditions also corrodes the outside of the exhaust system. If the tail pipe is rusted out, the muffler doesn't have much going for it either. Replace it. Because the exhaust pipe is nearer the engine, it doesn't need replacement as often. There is a gasket or sealing ring at the exhaust manifold that could burn out and have to be replaced. You can get by with regular tools when working on the exhaust system, but a cutting torch can save time and your knuckles. Various brackets hold the exhaust system in alignment, and rattles sometimes develop because of loose or broken mounts.

Raise the vehicle on stands and inspect the entire system by moving the parts around to check for loose connections and rattles. The old eyeball test should quickly spot rusted-out components. Check the suspected areas with a screwdriver. If the screwdriver pokes through,then the parts needed replacement. Check the exhaust manifold(s); they can come loose and the gaskets will burn out. A hissing sound indicates exhaust gas getting by.

You could plug the tail pipe with a potato and start the engine. The garage doors must be wide open. Check for escaping exhaust gas; it will be pretty evident. Remove the semibaked potato. The biggest frustration you will encounter is with the rusted bolts and joints. The joints are difficult to separate and the bolts break off. This is not so bad on the clamps. It is a good idea to replace them because they might break on assembly.

Breakage is particularly maddening when it happens at the exhaust pipe flange or the exhaust manifold. You then have to drill the broken piece out and retap it. Sometimes you have to remove more parts and chance breaking off more bolts or studs. Use penetrating oil or heat to keep this problem at a minimum.

To replace a tail pipe or muffler, first decide if you want to do the job. This is also dirty work. If you are not so keen on it, let a muffler shop at it. Remove the muffler clamps either by undoing them or twisting them off. Spray the joint or heat it up to break the rust and twist the parts loose.

A slitting chisel will help. If you cut the tail pipe off with a torch or hacksaw, you can peel away the short piece that is left. Make sure the muffler is good by tapping the outside of it with a wrench. Rattles or a dull sound indicate the need for replacement.

Order the parts using the proper information on body style, engine and model year. Don't compare with the old parts because they may not be the proper ones. Order clamps and brackets. Clean the joint surfaces on the remaining parts with sandpaper and wire brush. Slide the new parts in place. You will have to raise the rear of the body to get the tail pipe in place. Attach the brackets and align the parts so that nothing rubs or touches when the vehicle is in operation. Bounce it once or twice to check. Tighten all bolts and clamps. Start the engine and listen for any hissing sounds at the joints. Fix them now.

To replace the cross-over pipe on the V-8 models, soak the connecting bolts and nuts with penetrating oil and clean the exposed threads with a wire brush. Work the nuts back and forth at the exhaust manifold. Use more penetrating oil to free them up. Order gaskets and sealing rings with the cross-over pipe. Replace, align, and tighten securely. Start the engine and check for leaks. Fix if required.

To replace exhaust manifold gasket(s) that have burned out on the engine side, you must remove the exhaust manifold. Soak the connecting fasteners. On the six-cylinder models, the intake and exhaust manifolds are bolted together. Take them off as one piece. Replace the gasket and check that the intake manifold sleeves are in place. Tighten clamp bolts 15 to 20 foot pounds and the two end clamp bolts 25 to 30 foot pounds.

On the V-8 models, coat the end of the exhaust manifold gaskets around water passages and exhaust manifold bolts with a graphite type grease. Tighten bolts 25 to 35 foot pounds of torque. You should be able to replace these gaskets without disturbing the rest of the exhaust system.

To replace sealing ring(s) or gaskets on the exhaust-pipe side of the exhaust manifold is very similar to changing the cross-over pipe. Use plenty of penetrating oil to loosen up the fasteners. Replace the gaskets or sealing rings as required. Brackets support the exhaust system and insulate the natural exhaust vibrations from transferring to the chassis and passenger compartment. They all incorporate a flexible material that in time stretches or becomes brittle and breaks off. This allows the parts to rattle and weaken the other supporting joints. Replace the brackets by unbolting at the chassis and the exhaust system. If you find the parts of the exhaust system need replacing, do so now. It makes no sense to put new brackets on a rotting tail pipe.

SERVICING HEAT CONTROL VALVE

This valve is very simple in design but very important in smooth engine operation. The valve routes the exhaust gas through the intake manifold casting and preheats the manifold under warm-up and low-speed operation. This helps to vaporize the gasoline and prevents poor idle and lousy acceleration.

The value is normally held in closed position by a weight and a spring. The weight tries to hold the valve shut, but the pressure of the exhaust gas tends to open it. The spring is a thermostatic coil and opens the valve as the engine warms up and the speed increases. The combination of these two effects forces just enough exhaust through the intake manifold to provide the necessary vaporization of the gas.

The problem with the valve is that it can stick in the "open" or "closed" position. If it sticks in "closed" position the result will be poor engine performance, overheating, and detonation. If it sticks in "open" position poor fuel economy, poor engine performance, and vapor lock might be the result. Sticking is caused by short-distance driving or excessive oil consumption. The engine does not warm up and the condensation and carbon tend to seize the valve shaft in the housing.

Check the valve by trying to move the weight and the shaft. If it will move, start the engine, and as it heats up see if the valve opens completely. The tension of the spring is very important. If it is too tight, the heat will not be turned off the valve and it will be impossible to get a maximum fuel charge into the cylinder. This will result in a loss of power and maximum speed.

To service the valve, check the shaft to make sure it is free in the housing. If it is sticking, spray it with special manifold heat-valve lubricant. Do not use oil or it will burn and seize the shaft. Sometimes tapping the shaft back and forth will free it up. If it doesn't, then a new valve should be installed. This is more common on the V-8 models. To replace the valve, disconnect the right-hand exhaust pipe. The valve is located between the manifold and the exhaust pipe. Soak the fasteners with plenty of penetrating oil to prevent giving yourself more work than you want to do.

With the valve out, soak it in carburetor cleaner to loosen up the carbon. Work the valve back and forth until it is free. If the shaft is seized so tight that the valve will not move, tap the ends and keep applying carburetor cleaner. You can service this valve by replacing the problem parts or replace the entire valve.

If the shaft is free, check the thermostat spring tension. The spring should be wound up just enough to slip the outer end over the anchor pin. This is about a 1/2 turn of the spring from its position when unhooked. Replace the valve, use new gaskets, and don't break the fasteners on assembly. Start the engine and check that the valve operates properly. If the valve does not open fully, replace the spring and recheck.

On the six-cylinder models, the valve is located in the exhaust manifold. If it should seize and you cannot free it, then you will have to remove the manifolds and separate the intake from the exhaust manifold. Service by cleaning and replacement of parts if necessary. Use new gaskets and install the manifold back on the engine head. Start the engine and check that the valve opens fully.

Chapter 11

Sheet-Metal Repair

Just you and your car, and then you notice dust, air, and water leaks as you motor along. Dust and air leaks are annoying, but a water leak can lead to rusting problem. If not corrected a water leak will lead to expensive sheet-metal repairs. There is no repair for rust other than replacing the entire panel or cutting out the rusted part and replacing it with new material. Water leaks at the trunk, at the doors, or past any of the glass will be the start of a rust problem. We have to repair the leaks.

Winter driving is also the cause of rust because of the salt that is used for roads. Get in the habit of washing your car regularly during winter to remove the salt build-up. Wash under the fender wells and on top of the headlight buckets to remove any road dirt. Twice a year—spring and fall—pressure wash the entire under carriage to remove any material that will trap moisture. You might consider rustproofing the entire underbody to insulate the sheet metal from corrosion.

You might be able to correct some of the body leaks by adjusting the doors or the trunk lid. Other leaks can be fixed by replacing the weatherstrips. There are also adjustments for correcting the alignment of the front fenders and hood due to sagging. This is a problem caused by old age or accident.

The drivers door might need the replacement of the hinges to correct a sagging problem and a door closing problem can be fixed by adjusting the door latch striker plates. Windows that do not close properly will be the cause of leaks and wind noise. These are corrected by aligning the glass or replacing the window regulator system.

To replace doors or front fenders because of rust or damage due to an accident is simply a case of unbolting the old and replacing with the new. Chrome strips, mouldings, and name plates are replaced in a similar fashion. Small dents or creases can be repaired in a similar fashion. Small dents or creases can be repaired by the use of plastic body filler and spray paint. Rust out on the rear fenders or under the doors should be left to professional repairmen. The use of plastic body filler in rust areas is only a temporary repair, and you are fooling yourself if you think it is going to last. What you should want is a vehicle that is going to give many more trouble-free and enjoyable highway miles. The only way to reach that goal is with solid metal—not patches.

REPAIRING BODY LEAKS

You won't have any problem finding the large leaks, but those air whistles or drops of water are enough to drive you to a new car dealership. The new cars suffer from some of the same problems. You might find water inside the car after a car wash. This is a pretty good way to find leaks. Use a low water pressure and run the water along any of the sealing surfaces. Work from the bottom up, one door at a time, and keep checking or have a helper inside the vehicle. Do all the glass, bottom first, and then the sides and the top. Check the trunk see if water is getting in past the tailight housings or the rear window.

Damaged weather strips are the most common cause of leaks. If the weather stripping is dried out, cracked, or torn, it must be replaced. You might be able to fix a tear with silicone seal. To fix a windshield, seal clean the leaking area and apply silicone. If water is getting past the door window sealing strip, the strip will have to be replaced. Check the drain holes at the bottom of the doors. If you cannot purchase new weather strips, some silicone spray or a rubber preservative might help give them some new life and a door or trunk adjustment might help.

To remove the old weather strips, snap the clips out of the retaining holes and soften the old cement with lacquer thinner before you pull the weather strip off. Clean the mounting surface. Check the positioning of the weather strip before applying contact cement. On front doors, the formed bend fits in the cove area of the door.

The butt joint should be made at the bottom of the door only after the weather strip is in place. Cement the bottom of the weather strip and up a short distance on the door.

On front doors, an auxiliary weather strip is used at the hinge pillar to direct the water from the offset area to a drainage hole in the door pillar. To remove the old one, use a putty knife to snap out the two fasteners, soften the old cement, and remove the weather strip. When installing the new piece do not cover any of the drain hole. Use contact cement to hold the weather strip in its place.

To replace the weather strip on the trunk lid, remove the clips. Be careful not to distort them. Some models require only cement. The replacement weather strip is longer than required but do not cut it until you have the entire weather strip in place. Butt joint the two ends together and cement. While in the trunk area, check the clips that hold the mouldings to the rear fenders and silicone the holes they fit in. Check the rear window drainage.

If you are bothered with an air whistle when the car is moving, use masking tape to seal the window on the inside of the vehicle and peel back just enough to bring the noise out. If the air is leaking past the ventilator window gaskets and the gaskets are sound, then the ventilator window will need adjusting. If the air is coming past the door glass, the glass channels will need adjusting.

BODY ADJUSTMENTS

Most body adjustments align or position the various body panels. The doors should fit in their openings and the clearances remain parallel in relation to the rest of the body. The trunk lid should fit tight enough to seal with an even gap around the opening. The hood should fit flush with the front fenders and cowl. There is also an allowance made at the body mounting points on the frame for removing or adding shims to correct body alignment due to an accident. The front fenders are adjustable front and rear.

Before making the hinge or latch adjustment, open the door and see if it sags. Move the door up and down. If it moves, check the hinge for excessive wear. Replace the hinge(s). Remove the door striker plate and let the door hang free on its hinges when you are checking the alignment and spacing. To move the door up or down, or in and out, loosen the bolts attaching the hinge box to the body pillar. Push and pull the door into position and retighten the bolts. Check the alignment. The door should fit flush with the body and have about 1/4-of-an-inch gap around it.

To move the door backward or forward, you will have to loosen the hinge to door bolts. To do this, the trim pad must come off. To remove the inside handles, use a thin piece of metal inserted between handle and the trim pad. Push the spring clip out and place it back in the handle. Remove the moulding, the locking knob, and the arm rest. Remove the screws from the lower corners of the trim pad. Use a putty knife to pry the retaining clips from the slots in the inner door panel. Lift the trim pad up to free it from the retaining rail. Loosen the bolts attaching the hinge strap to the door and push or pull the door to even up the gap. Tighten the bolts. If the gap is fine but the door is tilted—meaning close at the top and wide at the bottom, or vice versa—then the body will have to be shimmed at the frame mounts. Shims can be added or subtracted to correct the body alignment.

You might find that the ventilator and door glass will need alignment. If there has been a wind whistle and you have tracked it down to the door ventilator, now is the time to fix it. To replace the gasket the ventilator must be removed. Remove the screws that hold the regulator and disengage it from the "T" shaft. Lower the regulator into the door and remove it through the small access opening. Loosen the upper glass run channel and remove the four attaching screws. Just below the door handle, there is an adjusting stud and nut that must be removed. Pull the ventilator inward and pull it out of the door.

Replace the gasket and secure the ventilator back in the door. If you are replacing the door glass, the procedure is similar to this point (except don't put the ventilator in). Remove the large access door and lower the glass into the door. Remove the two screws that hold the sash channel cam to the glass channel. Disengage the window from the sash channel cam, raise the glass almost to a fully closed position. Tilt inward and remove from the door. Replace the glass with one of similar size. If you haven't worked with glass, let the glass shop do it.

Why buy two windows when you only need one? Check the upper glass run channel, and if it worn out replace it. Work through the large access hole to get at the fasteners. While you have the door apart, check the window regulator and the lock mechanism. Don't replace with old parts unless you know for sure that they operate properly. It's a lot of work to get this far so do it right the first time.

Install the window glass first and then the ventilator. To adjust a glass that is slanted, loosen the two stationary cam screws that are located just under the lock connecting rod, adjust the rear of the cam up or down, and re-tighten the screws.

To adjust the ventilator glass, loosen the lock nut and turn the stud in or out and position the stud forward or backward to line the glass up. Tighten the glass run channel in proper position. Use silicone to seal the access doors and any of the bolts that were disturbed. All the inner panel openings should be sealed to prevent water leakage and damage to the trim pad. Replace the trim pad by aligning the clips with the holes in the inner door panel. Don't forget the inner springs on the regulator mechanisms. Replace the handles. If the springs are damaged replace them with new ones. The handles should be located as a mirror image to the other side. Replace the mouldings and any other parts that were removed from the door. Rear doors come apart in a similar way for service to glass and regulators.

To adjust the striker plate, loosen the three attaching screws. If the striker plate was removed for door alignment, replace it in position. If you are replacing the door, do a visual check to see if the lock extension will engage in the striker notch. The striker plate should be shimmed out so that inside face of the striker notch is 1/8 of an inch from the center of the lock extension. Longer screws are required with spacers. The door should close tightly without being forced.

To align the front fenders with the front doors when the gap is not parallel, remove or install shims at the fender to cowl attaching bolts. To move the bottom of the front fenders in or out, use shims at the body side rail bolt. Shims under the radiator support allow for an up or down movement or forward backward movement of the entire front sheet metal assembly.

To adjust the hood up or down, slotted holes are provided at the cowl attaching point. To allow for an even gap between hood and fenders, the hood hinges can be moved in or out. The tie bar can also be shimmed at the fender-attaching points—depending on which way the gap has to go. Try for even gaps and smooth metal alignment. Three stationary hood bumpers are located on each side at the fender-to-fender skirt-attaching flange. There are two adjustable bumpers on each side: one at the cowl and the other at the radiator grill reinforcement bar. The hood locking plate is adjusted by moving the plate on its slotted holes. There is no adjustment for hood spring tension. The springs must be replaced when the hood begins to drop down and you are trying to work it in the engine compartment.

Take it easy on this job and do it with safety. A coiled spring has a lot of energy when it is stretched. The trunk lid can be adjusted forward, rearward, or to either side by loosening the hinge strap bolts, moving the lid in the desired location, and retightening the bolts. Have a good fit to prevent dust and water from entering and rusting the trunk from the inside out. Keep this a clean area. It is not a holding compartment for junk. The more weight in here the less gas mileage you can expect.

You can adjust the amount of effort it takes to open or close the lid by moving the torsion rods into different positions on the hinge boxes. The top position is the one that springs the lid up the easiest but takes the most effort to close it. You can try different tensions by moving the right side and leaving the left side as is. To adjust the lid lock striker plate, loosen the attaching bolts and slide the striker plate on its beveled anchor plate. To check engagement pack some modeling clay into the hook of the latch. Slam the lid shut, open it, and measure the compressed clay. It should be to 5/32" thick. The trunk lid should close without excessive force and the striker plate should be centered in the latch.

REPLACEMENT METAL

The easiest metal to replace is the chrome, but you better see that the body metal is sound. Tail light and park light housings won't give you too much of a problem, and head light frames are a breeze. Hoods, doors, trunks and the front-end sheetmetal is strictly a matter of removing and replacing with NOS or good used metal. You might end up with a car of many colors but a paint job will cover all. Bumpers and grilles are easy to exchange. On any of these replacements, make good diagrams of where certain bolts and shims fit so that the assembly is a bit easier.

If you are replacing metal because of an accident, you will have to straighten out and align the mounting area or the replacement metal will not fit. Remove all the damaged metal. If the frame needs straightening have that done first. If this is a front-end accident, remove all the front-end sheet-metal from the doors forward. If the body has been squeezed a bit, have the body shop pull it back into alignment.

Side damage is a bit more diffcult. This is especially true if the roof and floor are buckled. Check with a body shop regarding cost if you want to retain this particular body style. Otherwise look for another body and switch over the chassis and running gear. Damage from the rear is probably best left to the body shops (if the rear quarter panels have to be replaced). A roll-over (if you're still around) is a total loss and you will have to go body hunting.

If you are repairing body rust, the only way to go is to cut out the rust to sound metal and weld in new panels. Align the panels carefully and remove all the rotten metal. This is a repair better left to body shops. If you can weld and have a welder, then have a go at it. Panels can be welded just about any where there is or was rust. You can form your own or purchase preformed ones. Some body work will have to be done to correct warpage, but keep the heat low and skip weld to keep warpage at a minimum.

Grind the weld flat and body work the patch if it doesn't fit in with the surrounding metal. Small imperfections could be filled with plastic filler. Treat this area as you would any bare metal it must be sanded smooth, primed and painted. You must match the primer to the type of paint that is on the car. A lacquer primer should be used on cars that have been painted with lacquer paint and an enamel primer should be used for enamel paint. Let the paint shop where you buy your supplies help you with this choice. The final coat of paint should be an exact match of the vehicle's color. You can have this mixed and put in a pressurized can so that you can apply it yourself. If you are doing a large area, rent a spray gun and a compressor to make the job easier and faster.

REPAIRING DENTS

The only skill and the most important one that you will need to repair dents is to know when you shouldn't attempt a certain repair. Once you pound a piece of metal out of shape, its going to take a very experienced bodyman many hours to bring it back to its original contour. In most cases, they might not want to take the job on after you have displayed your "skill."

Bodywork is *not* the pounding of sheet metal; it is the tapping of the dent to bring it back to its original shape. Once you overstretch the metal, a different series of work procedures has to be used. Work slowly and carefully and you will bring the dent out far enough so that a small amount of plastic filler will smoothen the surface.

Body hammers are lightweight striking tools that have contoured faces. You can get by with a small ball-peen hammer and tap the dent out while you support it with a larger hammer, but it is nice to have a few bodywork tools. A pick hammer and an all-purpose dolly will save a lot of time and make the work easier. A Surform tool to remove excess filler works faster than sandpaper. With a sanding block or board (depending on the size of the repaired area) and a face mask to filter out the plastic dust, you are ready to go to work.

Small dents might not need hammering out but filler should not be used more than 1/8 of an inch thick. It is a case of seeing what you have and bringing the metal up. If the dent or crease is any where near the chrome trim or light housings, remove as much as you can to make a better work area. Wash the area with water and then solvent to remove wax and grease. Wipe the solvent off before it dries. Examine the dent to see if you can get the dolly behind it to support the panel. If you strike the panel without a back-up support, a series of valleys will be formed. Areas that cannot be reached with the dolly will have to be pulled out. Clean the back of the panel where the dolly will have to be pulled out. Clean the back of the panel where the dolly will be held.

The hammer technique that you want to develop is one of slap-ping the metal with the hammer, using a glancing blow. You are not the village blacksmith. Do not hold the dolly directly behind the spot you are striking because the metal will stretch. Once you stretch the metal out of its normal shape, it will not stay flat and will tend to act like the bottom of a metal oil can. Take it easy. Support the back of the panel with the dolly and tap away, work-ing from the shallow part of the dent to the center. You can tap with the dolly and use the hammer to work down the crease.

Check for smoothness with your hand palm down, fingers to-gether. Run your entire hand over the surface of the work. You will be able to feel the high and low spots. Use the pick end of the body hammer to get in where the dolly won't reach and force the metal outward in a series of dimples. Use the dolly on the outside in this case. Keep checking for smoothness and contour. Use light strokes, move the dolly around, and the right shape will slowly develop. With the contour, run your hand over the area to find the low spots. Work them up. Better to stay just below rather than go over the original shape. Take your time you can't hurry any of this along.

Areas that you can't get the dolly behind or the pick hammer to will have to be pulled up. Body shops use a slide hammer for this, but you can get by with a 3/16-inch diameter sheet-metal screw. Use an electric drill and drill 1/8-inch holes along the length of the dent, starting 1 inch from the crease. Again you are working from the shallow end to the center. Insert the screw into the drilled hole and grasp it firmly with a pair of locking pliers. Pull straight out. Do not twist because the screw could tear out of the metal.

To help release the metal, tap the crease with the body hammer while pulling it out. Continue pulling and tapping until the dent is as close to the original shape as possible. You might have to drill a second row of holes if you can't get all the dent up. Always work toward the center and don't stretch the metal beyond its original shape. Keep checking for low spots by running your hand over the panel. If you detect a high spot, leave it alone for now.

Attach a grinding pad and grinding disc of 16 to 24 grit to an electric drill. Grind the surface clean of all paint and rust. Wear safety goggles. The disc scratches will provide an extra "tooth" or grip to which the plastic filler will bond. Grind an inch or two past the damaged area and keep the disc moving to prevent burning the metal. If there is any paint left between the creases, use a scratch awl to pick it out. Remove all traces of paint and get down to the bare metal.

Check the surface for smoothness by holding your hand flat and running it back and forth slowly over the entire area. Gently tap the low spots up using the dolly and hammer. Try to tap the high spots down using the dolly for support. Use a steel straightedge over the damaged area to give you a better idea of the high and low spots. If you are having difficulty lowering a high spot, then it means that the metal has stretched and you will have to shrink it back down. Get a piece of wood and shape the end of it so that the contour will fit the back of the panel. Hold the wood in place and use the pick hammer to drive the metal down. Check what you are doing and leave the contour as close to original as you can. Set the drilled holes down in a similar manner.

Read the instructions on the preparation of the plastic filler and then follow them. Stir the filler from the bottom up in order to mix the resin thoroughly and blend it in. Loosen the cap on the cream hardener and knead the tube to get a smooth paste. If the day is humid or the temperature is below 70 degrees Fahrenheit, the area to be repaired should be warmed up with a heat lamp. Mix the filler and hardener on a piece of glass or metal (not cardboard) until a uniform color is achieved. Use a rubbing motion for mixing (not a stirring one) because trapped air bubbles in the mixture will cause pinholes in the material after it has been applied.

Apply a thin coat to the prepared metal; use a good firm stroke. Then build up the low areas with succeeding coats. Work quickly because the mixture will start to set in about 15 minutes. As soon as the filler has a rubbery feel, use the Surform tool to shape the basic contour. If you wait too long, it will be too hard to work with the file.

Once the shape is obtained, use 80-grit, open-coat sandpaper to finish sanding the area to the correct contour. If you have a large area to sand, use a sanding board. A sanding block is fine for small areas. This is a dusty operation so a face mask should be used to filter out the plastic dust. Continue sanding until the filler is smooth, level, and flush with the metal.

Check the shape with your hand and tap down any high spots. Such spots are easy to find because the bare metal exposes them. You might need additional coats of filler to get the contour right. Blow the sanding dust off before you apply new coats of filler. Once the contour is smooth, apply a final coat of filler. This is a very thin coat and you will not be able to file it. Make it as smooth as you can and apply it in one direction. Let it dry and sand with 180-grit sandpaper. This is the final sanding before priming, so sand into the painted area and feather edge this paint. Check for smoothness with your hand. Sand in one direction. Where the sanding block won't fit, you will have to use only the paper. If there are a few pinholes or scratches, they will be filled later. The smoother the surface the better the finish.

Use masking tape and newspapers to cover the areas that will not be painted. Visit the local paint store and purchase some primer and wet-and-dry sandpaper. How much depends on how large the repair area is. If you can get by with a couple of spray cans that is great. The sandpaper should be 220 and 320 grit. Use light-colored primer under light paint and dark colored primer under dark paint. Have the shop mix the final paint color and put it in a spray can. You might be able to match a premix if the paint hasn't faded too much. Buy a tube or can of glazing putty to cover the small blemishes that are left in the repair area.

Wipe the area with a clean, dry rag. If any surface rust has formed, sand it off. Shake the spray can to mix the primer and spray on several light coats. Avoid sags and runs by keeping the can moving. Let the primer dry. Apply the glazing putty with a rubber squeegee. Use a very light coat until the entire area is covered. Work quickly because this putty dries fast. For best results, let the putty dry overnight. If you are working outside, the minimum temperature should be 60 degree Fahrenheit. Now you are ready for wet sanding. You can use a garden hose for water flow or a wet sponge. The water prevents the paper from clogging. The residue will stain concrete so get it off as soon as you can.

Start with the 220 paper and support it on a sanding block. In areas where you can't use the block, use your entire hand (not just the fingers or you will sand grooves into the surface). Avoid the highspots where the bare metal appears. Check for smoothness. If you have a lot of scratches, wet sand them out. Check the feather edges. There must be a smooth surface between paint, metal, and filler. Let the surface dry and prime it again. Let the primer dry. Wet sand with 320 paper. Watch the high spots as they have very little primer on them.

If you cut through, they must be spot primed. Paint will not stick to the bare metal. Let the primer dry and then use a sheet of used 320 to lightly dry sand the primed areas. If you feel so inclined, you can let the primer dry for a couple of weeks and then spray on the color coat. If this is the only repair, however, then you might as well do it now.

Wash the panel with wax and grease remover. Use two rags, one wet the other dry, and wipe off the panel before it dries. Spray the color coat on (using a back and forth motion). Don't stay in one spot or you will build up too much paint causing it to sag or run. Apply four or five coats. Allow a couple of minutes in between for drying. Let the paint dry at least 24 hours and then apply some polishing compound to blend the two finishes.

To repair rock chips and small rust spots that have not rusted through, you will have to sand and prime the area and then apply the color coat. If the vehicle has been painted a number of times, it takes a number of coats to build the surface of the paint up to the present level. You will never have a perfect color coat if the preparation underneath is poorly done. If you fill these blemishes with paint but do not prepare the surface, they will always show.

Appendix A

Troubleshooting

FRONT AXLE AND WHEEL ALIGNMENT

Symptom and Probable Cause	Probable Remedy

Hard Steering

a. Lack of lubrication.
b. Tight spherical joints.
c. Underinflated tires.
d. Improper front end alignment.
e. Improper steering gear adjustment.
f. Tie rod ends out of alignment.

a. Lubricate chassis and steering gear.
b. If not corrected by lubrication, replace joints.
c. Inflate tires to recommended pressure.
d. Adjust front end alignment.
e. Adjust steering gear.
f. Align tie rod ends with ball studs.

Front Wheel Shimmy

a. Underinflated tires.
b. Broken or loose wheel bearings.
c. Improper toe.
d. Worn spherical joints.
e. Improper caster.
f. Unbalanced wheels.
g. Steering gear loose.
h. Tie rod ball loose.

a. Inflate tires to recommended pressure.
b. Replace or adjust wheel bearings.
c. Adjust toe.
d. Replace joints.
e. Adjust caster.
f. Balance wheel and tire assemblies.
g. Adjust steering gear.
h. Replace tie rod end.

Road Wander

a. Underinflated tires.
b. Lack of lubrication.
c. Tight steering gear.
d. Improper toe-in.
e. Improper caster and camber.
f. Worn tie rod ends.
g. Loose relay rod.

a. Inflate tires to recommended pressure.
b. Lubricate chassis and steering gear.
c. Adjust steering gear.
d. Adjust toe-in.
e. Adjust caster and camber.
f. Replace tie rod ends.
g. Adjust relay rod joint.

Wheel tramp

a. Wheel assembly out of balance.
b. Blister or bump on tire.
c. Improper shock absorber action.

a. Clean wheel and balance assembly.
b. Replace or repair tire.
c. Replace shock absorber.

Excessive or Uneven Tire Wear

a. Underinflated tires.
b. Improper camber.
c. Improper caster.
d. Improper toe-in.
e. Wheels out of balance.
f. High speed cornering.
g. Brakes dragging.

a. Inflate tires to recommended pressure.
b. Adjust camber.
c. Adjust caster.
d. Adjust toe-in.
e. Balance wheels.
f. Instruct driver.
g. Adjust brakes.

REAR AXLE

Symptom and Probable Cause	Probable Remedy

Excessive Backlash

a. Loose wheel bolts.

    a. Tighten nuts securely. Make sure the tapered end of nut is toward wheel.

b. Worn universal joint.

    b. Replace or overhaul joint.

c. Loose propeller shaft to pinion splines.

    c. Replace worn parts.

d. Loose ring gear and pinion adjustment.

    d. Adjust ring gear and pinion.

e. Worn differential gears or case.

    e. Replace worn parts.

f. Worn axle shaft or differential gear splines.

    f. Replace worn parts.

Klunking Noise in Axle or Vehicle Weight Shifts From Side to Side on Turns

a. Excessive end play in axle shafts.

    a. Replace axle shaft bearings and or retainer.

Axle Noise on Drive

a. Ring gear and pinion adjustment too tight.

    a. Readjust ring gear and pinion.

b. Pinion bearings rough.

    b. Replace bearing and readjust ring gear and pinion.

Axle Noisy on Coast

a. Ring gear and pinion adjustment too loose.
b. Pinion bearings rough.

c. Excessive end play in pinion.

a. Readjust ring gear and pinion.
b. Replace bearing and readjust ring gear and pinion.
c. Adjust pinion bearings or replace bearings.

Axle Noisy on Both Drive and Coast

a. Pinion bearings rough.

b. Loose or damaged differential side bearings.
c. Damaged axle shaft bearing.
d. Worn universal joint.
e. Badly worn ring gear or pinion teeth.
f. Pinion too deep in ring gear.
g. Loose or worn wheel bearings.

a. Replace bearings and adjust ring gear and pinion.
b. Replace or adjust differential side bearings.
c. Replace bearing.
d. Replace worn parts.
e. Replace ring gear and pinion.
f. Adjust by shimming.
g. Replace wheel bearings.

Axle Lubricant Leaks

a. Axle shaft bearing seals leaking.

b. Pinion shaft oil seal leaking.
c. Differential carrier to housing gasket leaking.

a. Replace axle shaft bearing seals and "O" rings.
b. Replace pinion shaft oil seal.
c. Replace gasket.

PROPELLER SHAFT AND UNIVERSAL JOINTS

Symptom and Probable Cause	Probable Remedy

Excessive Vibration

a. Worn universal joint.
b. Bent propeller shaft.
c. Universal joint yoke bearings worn.

a. Replace worn parts.
b. Replace bent shaft.
c. Replace worn parts.

Excessive Backlash

a. Worn universal joints.
b. Worn drive shaft or joint splines.

a. Replace worn parts.
b. Replace worn parts.

FRAME AND SHOCK ABSORBERS

Symptom and Probable Cause	Probable Remedy

Hard Riding

a. Shock absorber broken or seized.

a. Disconnect shock absorber and test action, replace if necessary.

b. Rear spring leaf inserts worn.

b. Replace inserts.

c. Excessive tire pressure.

c. Check tire pressure, maintain at 24 pounds.

Vehicle Too Flexible

a. Faulty shock absorber.

a. Disconnect shock absorber and test action, replace if necessary.

Shock Absorber Noisy

a. Faulty shock absorber.

a. Disconnect shock absorber and test action, replace if necessary.

b. Loose dust tube on rear shock absorbers.

b. Replace shock absorber or refasten tube.

c. Improper grommet installation or loose retaining nuts.

c. Inspect and correct as necessary.

Leaks Fluid

a. Faulty shock absorber.

a. Replace shock absorber.

REAR AXLE AND SUSPENSION SPECIFICATIONS

Item	
Type	Semi-Floating
Gears	Hypoid
Ratio	
Overdrive	4.11:1
Conventional	3.70:1
Powerglide	3.55:1
Backlash	.005"-.008"
Pinion	
Mounting	Overhung
Adjustment	Shim and self locking pinion drive flange nut and compressible bearing spacer.
Thrust	Through taper roller, pinion brg's.
Bearings	
Front	Timken cone and roller
Rear	Timken cone and roller
Differential Type	Two Pinion, Hyatt Barrel

137

Axle Shaft
  Type   Wheel Drive Flange Integral with Shaft
  Minimum Diam. $\frac{63}{64}''$

Axle Shaft Bearings   Ball type

Drive Torque   Rear Springs

Lubricant Capacity   3½ Pints

SUSPENSION

Spring Type   Semi-elliptical
No. leaves   Four
Inserts   Wax-impregnated Fabric
Length   58''
Width   2''
Bushings   Rubber Type

PROPELLER SHAFT

Type   Open Hotchkiss
U Joints   Sealed Needle Bearing

BRAKE SYSTEM

Symptom and Probable Cause	Probable Remedy
Pedal Spongy	
a. Air in brake lines.	a. Bleed brakes.
All Brakes Drag	
a. Mineral oil in system.	a. Flush entire brake system and replace all rubber parts.
b. Improper pedal to push rod clearance.	b. Adjust clearance.
c. Compensating port in main cylinder restricted.	c. Overhaul main cylinder.
One Brake Drags	
a. Loose or damaged wheel bearings.	a. Adjust or replace wheel bearings.
b. Weak, broken or unhooked brake retractor spring.	b. Replace retractor spring.
c. Brake shoes adjusted too close to brake drum.	c. Correctly adjust brakes.
Excessive Pedal Travel	
a. Normal lining wear or improper shoe adjustment.	a. Adjust brakes.
b. Fluid low in main cylinder.	b. Fill main cylinder and bleed brakes.

Brake Pedal Applies Brakes but Pedal Gradually Goes to Floor Board

  a. External leaks.

  b. Main cylinder leaks past primary cup.

Brakes Uneven

  a. Grease on linings.

  b. Tires improperly inflated.

  c. Spring center bolt sheared and spring shifted on axle.

Excessive Pedal Pressure Required, Poor Brakes

  a. Grease, mud or water on linings.

  b. Full area of linings not contacting drums.

  c. Scored brake drums.

  a. Check main cylinder, lines and wheel cylinder for leaks and make necessary repairs.

  b. Overhaul main cylinder.

  a. Clean brake mechanism; replace lining and correct cause of grease getting on lining.

  b. Inflate tires to correct pressure.

  c. Replace center bolt and tighten "U" bolts securely.

  a. Remove drums—clean and dry linings or replace.

  b. Free up shoe linkage, sand linings or replace shoes.

  c. Turn drums and install new linings.

TROUBLE DIAGNOSIS

Brake troubles may be easily diagnosed if the complaint is understood. The trouble will always show up in one or more of the four ways listed below. Related parts of the power brake system should be checked before dismantling the power brake when a malfunctioning brake system is experienced.

1. Hard pedal feel caused by—

a. Glazed linings
b. Grease or brake fluid on the linings
c. Bound up brake pedal linkage
d. Sticking vacuum check valve
e. Collapsed vacuum hose
f. Plugged vacuum fittings
g. Leaking vacuum reserve tank
h. Internal vacuum hose loose or restricted
i. Jammed vacuum cylinder piston
j. Vacuum leaks in unit caused by loose piston plate screws
k. Faulty diaphragm rubber stop in reaction diaphragm
l. Faulty vacuum cylinder piston seal

2. Severe brakes may be caused by—

a. Grease or brake fluid on linings
b. Scored drums

c. Reaction diaphragm leakage
d. Broken counter-reaction spring
e. Restricted diaphragm passage
f. Sticking vacuum action (Do Not Oil)

3. Pedal goes to floor (or almost to floor), caused by—

a. Brakes require adjustment
b. Air in hydraulic system
c. Hydraulic leak in lines or at wheel cylinders
d. Fluid reservoir needs replenishing
e. Compensating valve leak
f. Hydraulic piston seal leak
g. Compensating port or outlet fitting seal leak

4. Brakes fail to release (or slow release) due to—

a. Brakes improperly adjusted
b. Bound up brake pedal linkage
c. Restricted air cleaner or passages
d. Excessive hydraulic seal friction
e. Compensator port plugged
f. Faulty residual check valve
g. Piston stroke interference
h. Sticky vacuum valve (Do Not Oil)
i. Broken piston return spring
j. Dry vacuum piston leather packing

141

ENGINES

Symptom and Probable Cause	Probable Remedy

LACK OF POWER

1. Poor Compression

a. Incorrect valve lash

b. Leaky valves

c. Valve stems or lifters sticking

d. Valve springs weak or broken

e. Valve timing incorrect

f. Leaking cylinder head gasket

g. Piston rings broken

h. Poor fits between pistons, rings and cylinders

a. Adjust valve lash according to instructions under "Valve Adjustment"

b. Remove cylinder head and grind valves

c. Free up or replace

d. Replace springs

e. Correct valve timing

f. Replace gasket

g. Replace rings

h. Overhaul engine

2. Ignition System Improperly Adjusted

a. Ignition not properly timed

b. Spark plugs faulty

c. Distributor points not set correctly

a. Set ignition according to instructions under "Engine Tune-Up"

b. Replace or clean, adjust and test spark plugs

c. Set distributor points and time engine

3. Lack of Fuel

a. Dirt or water in carburetor

b. Gas lines partly plugged

c. Dirt in gas tank

d. Air leaks in gas line

e. Fuel pump not functioning properly

a. Clean carburetor and fuel pump

b. Clean gas lines

c. Clean gas tank

d. Tighten and check gas lines

e. Replace or repair fuel pump

Symptom and Probable Cause

Probable Remedy

4. Carburetor Air Inlet Restricted

   a. Air cleaner dirty

   b. Carburetor choke partly closed

   a. Clean air cleaner

   b. Adjust or replace choke mechanism

5. Overheating

   a. Lack of water

   b. Fan belt loose

   c. Fan belt worn or oil soaked

   d. Thermostat sticking closed

   e. Water pump inoperative

   f. Cooling system clogged

   g. Incorrect ignition or valve timing

   h. Brakes dragging

   i. Improper grade and viscosity oil being used

   j. Fuel mixture too lean

   k. Valves improperly adjusted

   l. Defective ignition system

   m. Exhaust system partly restricted

   a. Refill system

   b. Adjust or replace

   c. Replace belt

   d. Replace thermostat

   e. Replace water pump

   f. Clean and reverse flush

   g. Retime engine

   h. Adjust brakes

   i. Change to correct oil

   j. Overhaul or adjust carburetor

   k. Adjust valves

   l. See "Engine Tune-Up"

   m. Clean or replace

6. Overcooling

   a. Thermostat holding open

   a. Replace thermostat

EXCESSIVE OIL CONSUMPTION

1. Leaking Oil

a. Oil pan drain plug loose — a. Tighten drain plug
b. Oil pan retainer bolts loose — b. Tighten oil pan bolts
c. Oil pan gaskets damaged — c. Replace pan gaskets
d. Timing gear cover loose or gasket damaged — d. Tighten cover bolts or replace gasket
e. Oil return from timing gear case to block restricted, causing leak at crankshaft fan pulley hub on six cylinder models — e. Remove oil pan and clean oil return passages
f. Rocker arm cover gaskets or, on six cylinder models, push rod cover damaged or loose — f. Tighten covers or replace gaskets
g. Fuel pump loose or gasket damaged — g. Tighten fuel pump or replace gasket
h. Rear main bearing leaking oil into clutch housing or flywheel housing — h. Adjust or replace main bearing or main bearing oil seal

2. Burning Oil

a. Broken piston rings — a. Replace rings
b. Rings not correctly seated to cylinder walls — b. Give sufficient time for rings to seat. Replace if necessary
c. Piston rings worn excessively or stuck in ring grooves — c. Replace rings
d. Piston ring oil return holes clogged with carbon — d. Replace rings

HARD STARTING

1 Slow Cranking

Symptom and Probable Cause	Probable Remedy
a. Heavy engine oil	a. Change to lighter oil
b. Partially discharged battery	b. Charge battery
c. Faulty or undercapacity battery	c. Replace battery
d. Poor battery connections	d. Clean and tighten or replace connections
e. Faulty starter switch	e. Replace switch
f. Faulty starting motor or drive	f. Overhaul starting motor
e. Excessive clearance between piston and cylinder wall due to wear or improper fitting	e. Fit new pistons
f. Cylinder walls scored, tapered or out-of-round	f. Recondition cylinders and fit new pistons

2. Ignition Trouble

Symptom and Probable Cause	Probable Remedy
a. Distributor points burned or corroded	a. Clean or replace points
b. Points improperly adjusted	b. Readjust points to .016", adjust new points to .019".
c. Spark plugs improperly gapped	c. Set plug gap at .035"
d. Spark plug wires loose and corroded in distributor cap	d. Clean wire and cap terminals

145

e. Loose connections in primary circuit | e. Tighten all connections in primary circuit
f. Series resistance in condenser circuit | f. Clean all connections in condenser circuit
g. Low capacity condenser | g. Install proper condenser
h. Ballast resister faulty or out of circuit | h. Inspect and correct

3. Engine Condition

a. Valves holding open | a. Adjust valves
b. Valves burned | b. Grind valves
c. Leaking manifold gasket | c. Tighten manifold bolts or replace gasket
d. Loose carburetor mounting | d. Tighten carburetor
e. Faulty pistons, rings or cylinders | e. See "Poor Compression"

4. Carburetion

a. Choke not operating properly | a. Adjust or repair choke mechanism
b. Throttle not set properly | b. Set throttle
c. Carburetor dirty and passages restricted | c. Overhaul carburetor

POPPING, SPITTING AND DETONATION

1. Overheated Intake Manifold

a. Manifold heat control spring not properly installed | a. Adjust according to instructions under "Engine Tune-Up"
b. Manifold heat control valve sticking | b. Free up heat control valve

146

2. **Ignition Trouble**

   a. Loose wiring connections
   b. Faulty wiring
   c. Faulty spark plugs

       a. Tighten all wire connections
       b. Replace faulty wiring
       c. Clean or replace and adjust plugs

3. **Carburetion**

   a. Lean combustion mixture
   b. Dirt in carburetor
   c. Restricted gas supply to carburetor
   d. Leaking carburetor or intake manifold gaskets

       a. Clean and adjust carburetor
       b. Clean carburetor
       c. Clean gas lines and check for restrictions
       d. Tighten carburetor to manifold and manifold to head bolts or replace gaskets

4. **Valves**

   a. Valves adjusted too tight
   b. Valves sticking

   c. Exhaust valves thin and heads overheating
   d. Weak valve springs
   e. Valves timed early

       a. Adjust valve lash
       b. Lubricate and free up. Grind valves if necessary
       c. Replace valves
       d. Replace valve springs
       e. Retime.

Symptom and Probable Cause	Probable Remedy
5. Cylinder Head	
a. Excessive carbon deposits in combustion chamber	a. Remove head and clean carbon
b. Cylinder head water passages partly clogged causing hot spot in combustion chamber	b. Remove cylinder head and clean water passages
c. Partly restricted exhaust ports in cylinder head	c. Remove cylinder head and clean exhaust ports
d. Cylinder head gasket blown between cylinders	d. Replace cylinder head gasket
6. Spark Plugs	
a. Spark plugs glazed	a. Clean or replace spark plugs
b. Wrong heat range plug being used	b. Change to correct spark plugs
7. Exhaust System	
a. Exhaust manifold or muffler restricted causing back pressure	a. Clean or replace manifold and muffler

ROUGH ENGINE IDLE

1. Carburetor

a. Improper idling adjustment   a. Adjust according to instructions

b. Carburetor float needle valve not seating   b. Clean or replace

2. Air Leaks

a. Carburetor to manifold heat insulator or gasket leaks

    a. Tighten carburetor to manifold bolts or replace heat insulator or gasket

b. Manifold to head gasket leaks

    b. Tighten manifold to head bolts or replace gaskets

c. Air leaks in windshield wiper vacuum line

    c. Check for leaks and repair

3. Valves

a. Improper lash adjustment   a. Check and adjust valves

b. Valves not seating properly   b. Grind valves

c. Valves loose in guides or bores   c. Condition valves

4. Cylinder Head

a. Cracks in exhaust ports   a. Replace cylinder head

b. Head gasket leaks   b. Replace cylinder head gasket

ENGINE MISSES ON ACCELERATION

1. Carburetion

a. Accelerating pump jet misadjusted plugged

    a. Overhaul carburetor or, on eight cylinder

or vapor vent ball in pump plunger not working   models, adjust pump travel.

  b. Lean fuel mixture   b. Overhaul carburetor

2. Ignition Trouble

a. Faulty spark plugs   a. Clean, adjust or replace plugs

b. Faulty ignition wiring   b. Replace faulty wiring

c. Improperly adjusted or faulty distributor points   c. Adjust or replace distributor points

d. Weak coil   d. Replace coil

Symptom and Probable Cause	Probable Remedy

3. Engine

a. Burned or improperly adjusted valves   a. Adjust, replace or grind valves

b. Leaky manifold gaskets   b. Tighten manifold or replace gaskets

c. Poor compression due to cylinder, piston or ring condition   c. Overhaul engine

d. Leaky cylinder head gasket   d. Replace gasket

ENGINE NOISE

1. Crankshaft Bearings Loose

a. Bearings improperly fitted   a. Readjust main bearings

b. Crankshaft journals out-of-round   b. Replace or recondition crankshaft

c. Crankshaft journals rough
d. Oil passages in block restricted
e. Insufficient oil
f. Improper grade and viscosity oil being used
g. Oil pump failure
h. Contaminated oil

2. Connecting Rod Bearings Loose

a. Worn bearings
b. Crankpins rough

c. Insufficient oil
d. Oil pump failure

e. Improper grade and viscosity of oil used

3. Pistons or Pins Loose

a. Excessive cylinder wear

b. Improperly fitted pistons or pins

---

c. Replace or recondition crankshaft.
d. Clean passages
e. Adjust or replace bearings. Replenish oil
f. Adjust bearings and change to correct oil
g. Replace oil pump, adjust or replace bearings and other damaged parts
h. Wash motor thoroughly. Adjust or replace bearings and other damaged parts

a. Replace bearings
b. Polish or replace shaft. Adjust or replace bearings
c. Adjust or replace bearings and replenish oil
d. Replace oil pump. Replace or adjust rod bearings
e. Replace rod bearings and change to proper oil

a. Hone cylinders and fit new pistons and rings. Make sure all abrasive that would cause cylinder wear is removed
b. Replace pistons or pins

c. Contaminated oil

d. Faulty fuel or ignition system causing unburned fuel to flush the oil from cylinder walls

e. Piston pin or bore wear

4. Engine Noise—General

a. Bent connecting rod

b. Excessive end play in camshaft on six cylinder models

c. Excessive crankshaft end play

d. Broken piston ring

e. Loose timing gears or chain

f. Dry push rod sockets

g. Bent oil gauge rod

h. Improperly adjusted valve lash

i. Sticking valves

c. Make necessary replacements, flush oiling system and use new oil

d. Make necessary repairs to fuel or ignition system, replace worn parts and change oil

e. Ream pin bore and install oversize piston pins on six cylinder models. Replace pistons and pins on eight cylinder models

a. Replace rod

b. Replace camshaft thrust plate, or correct end play by pressing gear on further

c. Replace main bearings

d. Replace broken ring and check condition of cylinder wall

e. Replace timing gears or chain

f. Polish and lubricate push rod sockets

g. Replace oil gauge rod

h. Adjust valve lash

i. Free or grind valves

FUEL SYSTEM

Symptom and Probable Cause	Probable Remedy
Excessive Fuel Consumption	a. Adjust idling mixture screws
a. Improper adjustment	b. Check and adjust float level
b. Improper float adjustment	c. Clean air cleaner
c. Dirty air cleaner	d. Check carburetor, fuel pump, fuel tank and connections for leaks
d. Fuel leaks	e. Check choke and throttle valve and manifold heat control valve or spring for proper operation
e. Sticking controls	
f. Improper engine temperature	f. Refer to Cooling Section
g. Dragging brakes	g. Refer to Brake Section
h. Engine improperly tuned	h. Tune engine—See Engine Section
i. Tires underinflated	i. Inflate to recommended pressure
j. Dirt in carburetor	j. Clean carburetor
k. Wrong jets	k. Install correct jets

153

Fast Idling

a. Improper adjustment
b. Controls Sticking
c. Automatic choke sticking
d. Poor automatic choke heat connection

    a. Adjust idling and throttle stop screws
    b. Free up controls and lubricate linkage
    c. Clean automatic choke
    d. Repair automatic choke heat connections

Engine Dies (Will Not Idle)

a. Idle speed or mixture screws improperly adjusted.
b. Low speed jet or idle passages plugged.
c. Vacuum leaks, carburetor or manifold
d. Float needle and seat loose, worn or sticking

    a. Adjust carburetor

    b. Disassemble and clean carburetor
    c. Replace necessary gaskets
    d. Replace needle and seat

Engine Misses on Acceleration

a. Accelerating pump jet plugged
b. Accelerating pump check valves sticking or leaking
c. Faulty power piston
d. Worn pump leather
e. Improper spark plug adjustment
f. Improper tappet adjustment
g. Sticking or burned valves

    a. Disassemble and clean carburetor
    b. Disassemble and clean carburetor

    c. Free-up or replace power piston
    d. Replace pump plunger
    e. Adjust spark plugs—See Engine Section
    f. Adjust tappets—See Engine Section
    g. Free up sticking valves or replace burned valves

FUEL SYSTEM SPECIFICATIONS

Air Cleaner

Make .... AC

Type:

Standard .... Mesh

Optional .... Oil Bath

Carburetor

Make and Type:

6-cylinder ....GM Model "BC" Downdraft
with Automatic Choke

8-cylinder .....GM Model 2GC Downdraft,
2 barrel with Automatic Choke

8-cylinder......Carter WCFB Downdraft, 4
(opt.)   barrel with Automatic Choke

Fuel Pump, 6-cylinder

Make, Model .... AC, model EM

Pressure .... 3½ to 4½ psi

Fuel Pump, 8-cylinder

Make, Model .... AC, model EM

Pressure ....4 to 5¼ psi

CLUTCH

Symptom and Probable Cause	Probable Remedy
Slipping	
a. Improper adjustment	a. Adjust pedal travel
b. Oil soaked	b. Install new disc
c. Worn splines on clutch gear	c. Replace transmission clutch gear
d. Lining torn loose from disc	d. Install new disc
Grabbing	
a. Oil on lining	a. Install new disc
b. Worn splines on clutch gear	b. Replace transmission clutch gear
c. Loose engine mountings	c. Tighten or replace mountings
Rattling	
a. Weak retracting springs	a. Replace springs
b. Throwout fork loose on ball stud	b. Check ball stud and retaining spring and replace if necessary
Noisy	
a. Worn throwout bearing	a. Replace bearing

Pedal Height, Tension

a. Pedal low, clutch lever not contacting rubber stop bumper.　　　**a.** Increase pull-back spring tension.

b. Pedal low, clutch lever compressing rubber stop bumper.　　　**b.** Adjust pedal stop.

c. Pedal high.　　　**c.** Adjust pedal stop.

d. Lack of free pedal travel feel.　　　**d.** Decrease pull-back spring tension.

CLUTCH SPECIFICATIONS

Type.................................Single Plate Dry Disc

Disc Diameter—V-8 ....10"

　　　　　　—Line 6 ....9½"

Clutch Pressure Spring

Type ....Diaphragm

Diameter ....9"

Clutch Release Bearing

Type ....Sealed Ball

Make ....New Departure

Clutch Pilot Bearing

Type.............Oil Impregnated Bushing

Clutch Pedal

Pedal Pressure..................6 to 9 Pounds

　　　　　　To Start Pedal Movement

COOLING SYSTEM

Symptom and Probable Cause	Probable Remedy
Overheating	
a. Lack of coolant	a. Refill system and check for leaks
b. Fan belt loose	b. Adjust
c. Fan belt oil soaked	c. Replace fan belt
d. Thermostat sticks closed	d. Replace thermostat
e. Water pump inoperative	e. Repair or replace water pump
f. Cooling system clogged	f. Clean system and reverse flush
g. Incorrect ignition timing	g. Retime engine
h. Brakes dragging severely	h. Adjust brakes
Overcooling	
a. Thermostat remains open	a. Replace thermostat
b. Extremely cold climate	b. Cover part of radiator area

Loss of Coolant

a. Leaking radiator
b. Loose or damaged hose connection
c. Leaking water pump
d. Loose or damaged heater hose
e. Leaking heater unit
f. Leak at cylinder head gasket

g. Cracked cylinder head
h. Cracked cylinder or block expansion plug loose

i. Engine operating at too high temperature

a. Replace or repair
b. Tighten or replace hose connections
c. Repair water pump
d. Tighten or replace hose
e. Replace or repair heater core
f. Replace gasket and tighten bolts securely and evenly
g. Replace cylinder head
h. Make necessary repairs or replacements

i. See overheating causes

Circulation System Noisy

a. Pump bearing rough
b. Fan blades loose or bent
c. Fan belt noisy in pulley

d. Fan belt inner plies loose
e. Improper fan to radiator clearance

a. Replace pump
b. Tighten or replace fan blades
c. Dress with belt dressing or soap and adjust
d. Replace fan belt
e. Adjust clearance, 5/8" to 3/4"

159

COOLING SYSTEM SPECIFICATIONS

Cooling System Capacity

All models without heater .... 16 Qt.
Add 1 quart for any model with heater.

Water Pump

Type and drive ...... Centrifugal by fan belt

Location .... Front of cylinder block

Capacity .... 6 cylinder, 55 gal. per minute at 4000 RPM engine speed.
8 cylinder 44.5 gal. per minute at 4000 RPM engine speed.

Impeller Location .... 6 cylinder—cyl. block
8 cylinder—pump body

Bearings .... Permanently lubricated and sealed ball.

Seal .... Molded rubber and fiber, automatically spring adjusted.

Fan

Diameter .... 17"
Number of Blades 4
Fan to Engine Speed Ratio .... .95:1

Fan Belt

Adjustment .... Movable generator

Deflection .... Light load midway between generator pulley and fan pulley.
6 cylinder— 5/16"
8 cylinder—13/16"

Thermostat .... Normal rating 160°F.
Starts to open 157°-163°
Fully open 183°F.

Location .... 6 cylinder head water outlet
8 cylinder intake manifold water outlet

Radiator Core

Make and Type ...... Harrison, Ribbed Cellular

Frontal Area—

	Standard Trans.	Powerglide
sq. in.		
6 cyl.	385	384.5
8 cyl.	357	354.6

(Powerglide models incorporate transmission oil cooler in radiator.)

BATTERY AND STARTING CIRCUIT

Symptom and Probable Cause	Probable Remedy

Slow Engine Cranking Speed

Partially discharged battery	Charge or change battery and determine cause of battery condition
Low capacity battery	Cycle battery to improve capacity or replace it
Faulty battery cell	Replace battery
Loose or corroded terminals	Clean and tighten terminals
Under capacity cables	Replace battery cables
Burned starter solenoid switch contacts	Replace solenoid
Internal starting motor trouble	Overhaul starting motor
Heavy oil or other engine trouble causing undue load	Make necessary repairs to engine

Starter Engages but Will Not Crank Engine

Partially discharged battery	Charge or change battery
Faulty battery cells	Replace battery
Bent armature shaft or damaged drive mechanism	Overhaul starter
Faulty armature or fields	Overhaul starter

Starter Will Not Run

Battery fully discharged	Replace or charge battery
Disconnected battery cables	Replace faulty cables
Shorted or open starter circuit	Make necessary repairs

GENERATING CIRCUIT

Low Charging Rate

Fully charged battery and low charging rate — This is a normal condition with a fully charged battery

Fan belt slipping — Replace or adjust belt

Generator commutator dirty — Clean commutator

High resistance in charging circuit — Check charging circuit progressively and make necessary repairs to remove high resistance

Too low voltage setting of voltage regulator unit — Adjust voltage regulator

Oxidized voltage regulator points — Clean and adjust points

Partially shorted field coils — Overhaul generator

High Charging Rate With Fully Charged Battery

Voltage regulator setting too high — Adjust voltage regulator

Voltage regulator points stuck — Clean and adjust points and readjust regulator

Regulator unit improperly grounded — Remove regulator and clean connections. Readjust regulator

Generator field circuit to regulator short circuited — Test to locate short circuit and make necessary repairs

Shunt field circuit short circuited within regulator — Replace regulator

Low Battery and No Charging Rate

Fan belt broken or loose
Replace or tighten fan belt

Charging circuit open between regulator and battery
Locate open circuit and make necessary repairs

Cut-out voltage winding open circuited
Replace regulator unit

Corroded points in current and voltage regulator
Clean points and readjust regulator

Open circuit between generator and regulator
Locate open circuit and make necessary repairs to wiring

Internal trouble in generator
Overhaul generator

IGNITION CIRCUIT

Engine Will Not Start
(See Starting and Fuel System Troubles)

Weak battery
Charge battery

Excessive moisture on high tension wiring or spark plugs
Dry parts

Cracked distributor cap
Replace cap

Faulty coil or condenser
Replace faulty unit

Coil to distributor high tension wire not in place
Properly install wire

Loose connections or broken wire in low tension circuit
Tighten or replace wires

Improperly adjusted or faulty distributor points
Clean and adjust or replace points

Hard Starting
(See Starting and Fuel System Troubles)

163

Faulty or improperly set spark plugs	Clean and adjust or replace spark plugs
Improperly adjusted or faulty distributor points	Clean or replace and adjust points
Loose connections in primary circuit	Tighten loose connections
Worn or oil soaked high tension wires	Replace high tension wires
Low capacity condenser	Replace condenser
Low capacity coil	Replace coil
Faulty distributor cap or rotor	Replace faulty part

Engine Misfires

Dirty or worn spark plugs	Clean or replace plugs
Damaged insulation on high tension wires or wires disconnected	Connect or replace wires
Distributor cap cracked	Replace cap
Poor cylinder compression	See Engine Troubles
Improper distributor point adjustment	Adjust distributor points

SPECIFICATIONS

BATTERY

Make	Delco-Remy
Plates per cell	9
Ampere hour capacity (at 20 hour rate)	50
Voltage	12
Specific gravity—fully charged	1.260-1.280 at 80°F

| | —¾ charged | 1.215 at 80°F |
| | —unsatisfactory | Below 1.215 at 80°F |

Maximum permissible specific gravity variation between cells with specific gravity over 1.215 — .025

GENERATOR

Make	Delco-Remy
Brush spring tension	28 oz.
Cold output	25 amps. at 14 volts and 2780 RPM
Field Current Draw	1.5-1.62 amperes at 12 volts
Current draw when run as motor	Average 4.7 amps, Max. 5.6 at 12 volts and 870-1070 RPM
Current draw at stall	Average 48.0 amps, Max. 58.0 at 12 volts and 0 RPM

REGULATOR

Make	Delco-Remy
Voltage regulator armature air gap	.075"
Voltage regulator setting	13.8-14.8 volts at 80°F
Current regulator armature air gap	.075"
Current regulator setting	23-27 amps.
Cut-Out relay closing voltage setting	11.8-13.5 volts
Cut-out relay points open (reverse flow)	No specifications—points must open

| Cut-out relay armature air gap | .020" |
| Cut-Out relay point opening | .020" |

STARTING MOTOR

Make	Delco-Remy
Brush spring tension	35 oz. Minimum
Free Speed	
Volts	10.3
Amperes	75 Maximum
RPM	6900 Minimum
Solenoid	
Hold-in winding	18-20 amps at 10 volts
Both windings	72-76 amps at 10 volts

DISTRIBUTOR

Make	Delco-Remy
Type of Advance	Centrifugal and Vacuum
Breaker Point Gap	New .019" Used .016"
Stabilizing Spring Tension (8 cylinder)	18-24 oz.
Friction Between Plates (8 Cylinder)	15 oz. Max.
Breaker Arm Spring Tension	19-23 oz.
Condenser Capacity	.18-.23 microfarad
Cam Angle (Dwell)	26-33 Degrees

Rotation	Clockwise (viewed from top in installed position)
Firing Order—6 Cylinder	1-5-3-6-2-4
—8 Cylinder	1-8-4-3-6-5-7-2
Ignition Timing—6 Cylinder	T.D.C.
—8 Cylinder	4° B.T.D.C.
Centrifugal Advance—6 Cylinder	
Start	0-2 Degrees @ 375 Distributor RPM
Intermediate	4-6 Degrees @ 700 Distributor RPM
Intermediate	9-11 Degrees @ 1350 Distributor RPM
Maximum	12-14 Degrees @ 1750 Distributor RPM
Centrifugal Advance—8 Cylinder	
Start	0-2 Degrees @ 400 Distributor RPM
Intermediate	8-10 Degrees @ 1150 Distributor RPM
Maximum	15-17 Degrees @ 1800 Distributor RPM
Vacuum Advance—6 Cylinder	
Start	4-6 "Hg.
Full Advance	7.5-10 "Hg.
Maximum Advance (Dist. Degrees)	6.5-8.5
Vacuum Advance—8 Cylinder	
Start	5-7 "Hg.
Full Advance	13.5-15.5 "Hg.
Maximum Advance (Dist. Degrees)	13-15

IGNITION COIL

Make	Delco-Remy

IGNITION RESISTOR

Make   Delco-Remy
Resistance   1.40-1.65 ohms

SPARK PLUGS

Make   AC
Type   AC-44-5 (Original Equipment and Service)
            AC-46-5 (Hotter Plug for Continuous City Operation—Service Only)
            AC-43-5 COM (Colder Plug for Continuous Heavy Duty Work—Service Only)

Size   14mm
Plug gap   .035"
Recommended Torque   20 to 25 ft. lbs.

168

SYNCHRO-MESH TRANSMISSION

Symptom and Probable Cause	Probable Remedy

Slips Out of High Gear

a. Transmission loose on clutch housing.
b. Dirt between transmission case and clutch housing.
c. Misalignment of transmission.

d. Clutch gear bearing retainer broken or loose.

e. Damaged mainshaft pilot bearing.
f. Shifter lock spring weak.
g. Clutch gear or second and third speed clutch improperly mated.

a. Tighten mounting bolts.
b. Clean mating surfaces.

c. Shim between transmission case and clutch housing.
d. Tighten or replace clutch gear bearing retainer.
e. Replace pilot bearing.
f. Replace spring.
g. Replace clutch gear and second and third speed clutch.

Slips Out of Low and/or Reverse

a. Worn first and reverse sliding gear.
b. Worn countergear bearings.
c. Worn reverse idler gear.
d. Shifter lock spring weak or broken.
e. Improperly adjusted linkage.

a. Replace worn gear.
b. Replace countergear bearings and shaft.
c. Replace idler gear.
d. Replace spring.
e. Adjust linkage.

Noisy in All Gears

a. Insufficient lubricant.
b. Worn countergear bearings.
c. Worn or damaged clutch gear and countershaft drive gear.
d. Damaged clutch gear or mainshaft ball bearings.
e. Damaged speedometer gears.

a. Fill to correct level.
b. Replace countergear bearings and shaft.
c. Replace worn or damaged gears.
d. Replace damaged bearings.
e. Replace damaged gears.

Noisy in High Gear

a. Damaged clutch gear bearing.
b. Damaged mainshaft bearing.
c. Damaged speedometer gears.

a. Replace damaged bearing.
b. Replace damaged bearing.
c. Replace speedometer gears.

Noisy in Neutral with Engine Running

a. Damaged clutch gear bearing.
b. Damaged mainshaft bearing.

a. Replace damaged bearing.
b. Replace damaged bearing.

Noisy in All Reduction Gears

a. Insufficient lubricant.
b. Worn or damaged clutch gear or counter drive gear.

a. Fill to correct level.
b. Replace faulty or damaged gears.

Symptom and Probable Cause	Probable Remedy
Noisy in Second Only	
a. Damaged or worn second speed constant mesh gears.	a. Replace damaged gears.
b. Worn or damaged countergear rear bearings.	b. Replace counter gear bearings and shaft.
Noisy in Low and Reverse Only	
a. Worn or damaged first and reverse sliding gear.	a. Replace worn gear.
b. Damaged or worn low and reverse countergear	b. Replace countergear assembly.
Noisy in Reverse Only	
a. Worn or damaged reverse idler.	a. Replace reverse idler.
b. Worn reverse idler bushings.	b. Replace reverse idler.
c. Damaged or worn reverse countergear.	c. Replace countergear assembly.

171

Excessive Backlash in Second Only

a. Second speed gear thrustwasher worn.
b. Mainshaft rear bearing not properly installed in case.
c. Universal joint retaining bolt loose.
d. Worn countergear rear bearing.

a. Replace thrustwasher.
b. Replace bearing, lock or case as necessary.
c. Tighten bolt.
d. Replace countergear bearings and shaft.

Excessive Backlash in All Reduction Gears

a. Worn countergear bushings.
b. Excessive end play in countergear.

a. Replace countergear.
b. Replace countergear thrustwashers.

Leaks Lubricant

a. Excessive amount of lubricant in transmission.
b. Loose or broken clutch gear bearing retainer.
c. Clutch gear bearing retainer gasket damaged.
d. Cover loose or gasket damaged.
e. Operating shaft seal leaks.
f. Idler shaft expansion plugs loose.
g. Countershaft loose in case.

a. Drain to correct level.
b. Tighten or replace retainer.
c. Replace gasket.
d. Tighten cover or replace gasket.
e. Replace operating shaft seal.
f. Replace expansion plugs.
g. Replace case.

172

TRANSMISSION SPECIFICATIONS

Type

Selective Synchromesh

Speeds

Three forward—one reverse.

Location

In unit with engine

Gears—Type

All helical

Bearings

Clutch Gear .... Ball Bearing
Countershaft .... 50 Rollers—⅛" Dia. x ¾"

Mainshaft

Front Pilot ...... 14 Rollers—3⁄16" dia. x 3⁄64"
Rear Pilot .... 24 Rollers—⅛" dia. x ½"
Mainshaft Rear .... Ball Bearing
Reverse Idler Bushing (front and rear) Bronze

Gear Ratio

First .... 2.94 to 1
Second .... 1.68 to 1
Third .... 1.00 to 1
Reverse .... 2.94 to 1

Service Data

Mainshaft Rear Bearing End Play .. .003" max.
Reverse Idler Gear Bushing
  Clearance .... .002"-.004"
Second Speed Gear Endplay ... Approx. .010"

TORQUE SPECIFICATIONS

Clutch Gear Bearing Retainer Cap
Screws .... 10-15 ft. lbs.
Side Cover Retaining Cap Screws .. 15-18 ft. lbs.

173

STEERING

Symptom and Probable Cause

Hard Steering

a. Lack of lubrication.

b. Pitman shaft to relay rod ball joint too tight.
c. Underinflated tires.
d. Improper adjustment.
e. Interference between steering shaft and mast jacket assembly caused by misalignment, bent steering shaft or damaged parts within the mast jacket assembly.

Loose Steering

a. Improper adjustments.
b. Loose pitman shaft to relay rod ball joint.
c. Worn steering knuckle ball joints.
d. Worn pitman shaft bushings.

Probable Remedy

a. Lubricate steering gear, tie rod ends, steering relay rod ball joints and steering knuckle joints.
b. Readjust ball joint.

c. Inflate tires to recommended pressure.
d. Adjust according to instruction.
e. Adjust or replace parts as required.

a. Adjust according to instructions.
b. Adjust ball joint.
c. Replace steering knuckle ball joints.
d. Replace bushings.

WHEELS AND TIRES

Symptom and Probable Cause	Probable Remedy

Front Wheel Shimmy

a. Loose wheel lugs.
b. Loose or broken wheel bearing.

c. Bent wheel.
d. Improper alignment.
e. Wheel out-of-balance.
f. Loose tie rod ends.

a. Tighten lugs.
b. Tighten or replace bearing and adjust according to instructions.
c. Replace or straighten wheel.
d. Front end alignment as per specifications.
e. Balance wheel.
f. Replace tie rod ends.

Hard Steering

a. Low air pressure in tires.
b. Lack of Lubrication.
c. Improper wheel alignment.
d. Sagging front or rear spring.
e. Bent wheel or spindle.
f. Broken wheel bearings.

a. Inflate tires to recommended pressure.
b. Lubricate according to instructions.
c. Front alignment correction.
d. Replace springs as required.
e. Straighten or replace wheel or replace spindle.
f. Replace necessary bearings.

Improper Tire Wear

a. Improper air pressures.
b. Not rotating tires as recommended.
c. Improper acting brakes.
d. Improper alignment.
e. High speed driving on turns.
f. Rapid stopping.

a. Inflate tires to recommended pressure.
b. Rotate tires according to instructions.
c. Correct brakes as required.
d. Align front end as per specifications.
e. Take turns more slowly.
f. Apply brakes slowly on approaching stop.

Noise in Front or Rear Wheels

a. Loose wheel lugs.
b. Broken or loose brake shoe return springs.
c. Broken or rough wheel bearings.
d. Scored drums.
e. Lack of lubrication.

a. Tighten wheel lugs.
b. Replace return springs.
c. Replace bearings according to instructions.
d. Replace brake lining and machine drums.
e. Lubricate as per instructions.

Loss of Air

a. Puncture in tire.
b. Faulty valve or valve core.
c. Rim defect.

a. Repair puncture.
b. Replace valve assembly or core.
c. Correct rim defect.

HEADLAMP AND CIRCUIT

Symptom and Probable Cause	Probable Remedy
Headlights Dim (engine idling or shut off)	
Partly discharged battery	Charge battery
Defective cells in battery	Replace battery
High resistance in light circuit	Check headlight circuit including ground connection. Make necessary repairs
Faulty bulbs	Replace bulbs or sealed beam units
Headlights Dim (engine running above idle)	
High resistance in lighting circuit	Check lighting circuit including ground connection. Make necessary repairs
Faulty bulbs or reflectors	Replace sealed beam units
Faulty voltage control unit	Test voltage control and generator. Make necessary repairs
Lights Flicker	
Loose connections or damaged wires in lighting circuit	Tighten connections and check for damaged wiring
Light wiring insulation damaged producing momentary short	Check light wiring and replace or tape damaged wires

177

Lights Burn Out Frequently

High voltage regulator setting — Adjust voltage regulator

Loose connections in lighting circuit — Check circuit for loose connections

Lights Will Not Light

Discharged battery — Recharge battery and correct cause

Loose connections in lighting circuit — Tighten connections

Burned out bulbs — Replace bulbs or sealed beam unit

Open or corroded contacts in lighting switch — Replace lighting switch

Open or corroded contacts in dimmer switch — Replace dimmer switch

Thermal Circuit Breaker Causing Current Interruption

Short in wiring — Check wiring of circuits in use for short circuits and make necessary repairs

Short within some light or instrument in use — Check lights or instruments for short. Headlamps and parking lamps are on separate circuit breaker from remainder of lighting units.

GASOLINE GAUGE

Symptom and Probable Cause	Probable Remedy
Gauge Shows Empty at All Times	
Tank unit shorted	Replace unit
Wire from dash unit to tank unit shorted	Replace wire or repair short
Float stuck in empty position	Replace tank unit
Dash unit improperly grounded on instrument panel	Properly ground dash unit
Gauge Shows Full at All Times	
Tank unit burned out	Replace tank unit
Wire between units disconnected or broken	Connect or replace wire
High resistance in wire between units	Clean connections and terminals
Float stuck in full position	Replace tank unit
Gauge Does Not Register Accurately (within normal limits)	
Bent hand on dash unit	Replace unit or straighten hand
High resistance in circuit	Check and correct circuit
Partial short in circuit	Correct cause of short
Loose electrical connections	Tighten connections at dash unit and tank unit

179

STOPLIGHT AND CIRCUIT

Will Not Light

Switch faulty — Replace switch
Wires broken, disconnected or loose — Make necessary repairs
Bulb burned out — Replace bulb
Loose connection or poorly grounded lamp body — Tighten loose connection or properly ground lamp body

HORNS

Will Not Blow

Loose connections or broken wire — Tighten loose connection or replace broken wire
Horn button not making contact — Adjust horn button contact
Horn improperly adjusted or faulty — Adjust or replace horn
Defective horn relay — Adjust or replace relay

Horn Tone Poor

Horn improperly adjusted — Adjust horn

Horn Operates Intermittently

Loose connections or intermittent connections in horn relay or horn circuit — Correct loose or intermittent connection condition or replace horn relay
Defective horn switch — Adjust horn button contact
Defective horn relay — Adjust or replace relay
Defects within the horn — Adjust or replace horn

GENERATOR TELLTALE LIGHT

Ignition on, Engine Not Running, Telltale Off

Indicator bulb burned out   Replace bulb
Open circuit or loose connection in telltale circuit   Locate open circuit or loose connection and correct

Telltale Light Stays On After Engine is Started

If on at idle only, improper idle speed   Adjust idle speed
Low generator output   Check generator output

Appendix B

Transmission Diagnosis

Any one of the following general complaints may be due to non-standard mechanical conditions in the overdrive unit:

1. Does not drive unless locked up manually.
2. Does not engage, or lock-up does not release.
3. Engages with a severe jolt, or noise.
4. Free-wheels at speeds over 30 mph.

These troubles may be diagnosed and remedied as described in the following paragraphs.

1. **Does not drive unless locked up manually.**

   a. Occasionally, the unit may not drive the car forward in direct drive, unless locked up by pulling the dash control. This may be caused by one or more broken rollers in the roller clutch, the remedy for which is the replacement of the entire set of rollers.

   b. This may also be caused by sticking of the roller retainer upon the cam. This retainer must move freely to push the rollers into engaging position, under the pressure of the two actuating springs.

   c. Sometimes this is due to slight indentations, worn in the cam faces by the rollers spinning, remedied by replacement of the cam.

2. **Does not engage, or lock-up does not release.**

   a. Dash control improperly connected—Unless the overdrive dash control wire is connected to the lockup lever on the left side of the overdrive housing in such a manner as to move the lever all the way back when the dash control knob is pushed in, it may hold the shift rail in such a position as to interlock the pawl against full engagement resulting in a buzzing noise when overdrive engagement is attempted.

      To correctly make this connection, loosen binding post at lever, pull dash control knob out ¼ in., move lever all the way to the rear, and tighten binding post.

   b. Transmission and overdrive improperly aligned — The same symptoms as above may also result from misalignment, at assembly, of the overdrive housing to the transmission case, resulting in binding of the overdrive shift rail, so that the retractor spring cannot move the rail fully forward, when the dash control knob is pushed in, and the transmission is not in reverse. Under such conditions, the unit may remain fully locked up.

      To test for this, be sure that the transmission is not in reverse; disconnect the dash control wire from the lockup lever, and feel the lever for free forward movement. If the lever can be moved forward more than ¼ in., it indicates that misalignment probably exists. To correct this, loosen the capscrews between the overdrive housing and transmission case, and tap the adapter plate and overdrive housing until a position is found where the rail shifts freely; tighten capscrews.

   c. Kickdown switch improperly adjusted — The position of the kickdown switch should be adjusted, by means of the two large nuts which clamp the switch shank, so the switch plunger travels $\frac{3}{16}$ in. before the throttle lever touches its stop.

Occasionally the large nuts which clamp the switch through the switch bracket are tightened sufficiently to bend the switch shank, thus preventing free motion of the switch stem. This may usually be remedied by loosening the upper of the two nuts.

d. Improper installation of solenoid—If car cannot be rolled backward under any circumstances and there is no relay click when the ignition is turned on, it probably indicates that the solenoid has been installed directly, without twisting into the bayonet lock between solenoid stem and pawl, thus jamming the pawl permanently into overdrive engagement. If the car will occasionally roll backwards, but not always, (and there is no relay click when the ignition switch is turned on) it may indicate that, upon installation, the bayonet lock was caught, and the solenoid forcibly twisted into alignment with the attaching flange, thus shearing off the internal keying of the solenoid. Under these circumstances, the end of the solenoid stem may not catch in the pawl, and upon release of the solenoid, the pawl will not be withdrawn promptly from engagement, but may simply drift out. If the solenoid stem end has its two flats exactly facing the two solenoid flange holes, it will not withdraw the pawl properly. If the stem can be rotated when grasped by a pair of pliers, it indicates that the internal keying has been sheared.

e. Improper positioning of blocker ring — Occasionally, either in assembly at the factory, or in service operations in the field, the internal parts of the overdrive unit may have been rotated with the solenoid pawl removed, causing the blocker ring to rotate, so that its two lugs are not located with respect to the pawl as shown in figs. 24 and 38. In other words, the solid portion of the blocker ring may be in alignment with the pawl, which will prevent full engagement of the pawl with the sun gear control plate.

To test for this condition, remove solenoid cover, pull dash control knob'out, roll car 2 ft. forward. Push dash control in, turn ignition switch on. Then ground the "KD" terminal of relay, and watch movement of center stem of solenoid. It should not move more than ⅛ in. when the solenoid clicks. Then, with the relay terminal still grounded, shift into low gear, and roll car forward by hand. Solenoid stem should then move an additional ⅜ in., as the pawl engages fully. These two tests indicate proper blocker action. Unless both tests are met, the blocker ring is probably not in the correct position.

3. **Engages with a severe jolt or noise.**

Insufficient blocker ring friction may cause the ring to lose its grip on the hub of the sun gear control plate. Check the fit and tension of the ring as described under "Cleaning and Inspection."

4. **Free-wheels at speeds over 30 mph.**

If cam roller retainer spring tension is weak the unit will free-wheel at all times. Check spring action as described under "Cleaning and Inspection."

ELECTRICAL

Any one of the following general complaints may be due to electrical trouble in the overdrive circuit.
1. Does not engage.
2. Does not release.
3. Does not kickdown from overdrive.

These troubles may be traced and remedied as described in the following paragraphs.

1. **Does not engage.**

   a. With the ignition switch on, ground the "KD" terminal of the solenoid relay with a jumper lead. If the solenoid clicks, the relay and solenoid circuits are in operating condition. If no click is heard in the relay, check the fuse and replace if defective.

   b. If the fuse is good, use a second jumper lead to connect the "SOL" and "BAT" terminals of the relay. If a click is now heard in the solenoid, the relay is probably at fault and should be repaired or replaced.

   c. If the solenoid does not click in Step b, check the wiring to the No. 4 terminal of the solenoid and replace if necessary. If the wiring is not defective, the trouble is probably in the solenoid. Remove the solenoid cover, examine the solenoid contacts in series with the pull-in winding and clean if necessary. Test again for clicks, as in Step b, after replacing solenoid cover and lead wires. Replace the solenoid if trouble has not been corrected.

   d. If the relay and solenoid circuits are in good condition as determined in Step a, leave the ignition switch on and make sure the manual control knob is in the overdrive position. Ground one and then the other of the two terminals next to the stem of the kickdown switch (identified as "SW" and "REL"). If the solenoid clicks when one terminal is grounded but not the other, replace the switch. If the solenoid does not click when either of the terminals is grounded, check the wiring between the relay and the kickdown switch and replace if defective.

   e. If the solenoid clicks as each terminal is grounded in Step d, ground the governor switch terminal. If the solenoid clicks, the governor switch may be defective. If the solenoid does not click, check the wiring between the kickdown and governor switches and replace if necessary.

2. **Does not release**

    a. Remove the connection to the "KD" terminal of the relay. If this releases overdrive, look for a grounded control circuit between the relay and governor switch.

    b. If the overdrive is not released in Step a, disconnect the lead to the "SOL" terminal of relay. If this releases the overdrive, replace the relay.

3. **Does not kickdown from overdrive**

    a. With the engine running, connect a jumper lead between the No. 6 terminal of the solenoid and ground. Operate the kickdown switch by hand. This should stop the engine. If it does, the solenoid is probably defective and it should be checked for dirty ground-out contacts or other defects within the ground-out circuit of the solenoid (fig. 28). Clean the contacts or replace the contact plate as required.

    b. If the engine does not stop in Step a, ground one and then the other of the two terminals (Identified as "IGN" and "SOL") farthest from the stem of the kickdown switch. The engine should stop when one of the two terminals (IGN) is grounded. If the engine does not stop when either of the terminals is grounded, the wiring or connections to the switch between the switch and coil are defective. When the other terminal (SOL) is grounded, the engine should stop when the kickdown switch is operated. If the engine does not stop when the kickdown switch is operated with the second terminal grounded, the kickdown switch is defective. If the trouble is in the kickdown switch, adjust the linkage to give more travel of the switch rod. If this does not correct the trouble, replace the kickdown switch.

If the kickdown switch operates as it should, check for an open circuit in the wiring between the kickdown switch and the No. 6 terminal of the solenoid.

c. If the trouble is not located by the above checks, the upper contacts of the kickdown switch may not be opening. To check for this condition, ground the overdrive control circuit at the governor switch. This should cause the solenoid to click. Operate the kickdown switch by hand. This should cause a second click as the solenoid releases. If there is no second click, adjust the linkage to give more travel of the switch rod. If this does not correct the trouble, replace the kickdown switch.

Index